AF567030

MICROTECHNOLOGY AND MEMS

Springer
Berlin
Heidelberg
New York
Hong Kong
London
Milan
Paris
Tokyo

MICROTECHNOLOGY AND MEMS

Series Editor: H. Baltes D. Liepmann

The series Microtechnology and MEMS comprises text books, monographs, and state-of-the-art reports in the very active field of microsystems and microtechnology. Written by leading physicists and engineers, the books describe the basic science, device design, and applications. They will appeal to researchers, engineers, and advanced students.

Mechanical Microsensors
By M. Elwenspoek and R. Wiegerink

CMOS Cantilever Sensor Systems
Atomic Force Microscopy and Gas Sensing Applications
By D. Lange, O. Brand, and H. Baltes

Micromachines as Tools for Nanotechnology
Editor: H. Fujita

H. Fujita (Ed.)

Micromachines as Tools for Nanotechnology

With 183 Figures

Springer

Professor Hiroyuki Fujita
The University of Tokyo
Institute of Industrial Science
4-6-1 Komaba, Meguro-ku
Tokyo 153-8505, Japan
E-mail: fujita@iis.u-tokyo.ac.jp

Series Editors:

Professor Dr. H. Baltes
ETH Zürich, Physical Electronics Laboratory
ETH Hoenggerberg, HPT-H6, 8093 Zürich, Switzerland

Professor Dr. Dorian Liepmann
University of California, Department of Bioengineering
466 Evans Hall, #1762, Berkeley, CA 94720-1762, USA

ISSN 1615-8326
ISBN 3-540-00856-X Springer-Verlag Berlin Heidelberg New York

Library of Congress Cataloging-in-Publication Data

Fujita, Hiroyuki, 1952-
Mircromachines as tools for nanotechnology / H. Fujita.
p. cm. -- (Mircotechnology and MEMS)
Includes bibliographical references and index.
ISBN 3-540-00856-X (alk. paper)
1. Nanotechnology. 2. Mircoelectromechanical systems. I. Title. II. Series.

T174.7.F85 2003
620'.5--dc21 2003045497

Springer-Verlag Berlin Heidelberg New York
a member of Springer Science+Business Media

http://www.springer.de

Typesetting: Camera-ready copy by the editor and LE-TEX GbR Leipzig using a Springer LATEXmacro
Production: LE-TEX Jelonek, Schmidt & Vöckler GbR, Leipzig
Cover concept: eStudio Calamar Steinen
Cover production: *design & production* GmbH, Heidelberg

Printed on acid-free paper SPIN 11009818 57/3111/YL - 5 4 3 2 1

Preface

Nanotechnology is the key technology of the 21st century that promises to bring dramatic new developments in electronics, communication networks, biotechnology, medical science and environmental research. The technology is expected to play an essential part in the life of society in the future. Governments around the world fund and promote major projects for research in nanotechnology while intensive efforts are also made in the private sector.

Nanotechnology deals with materials, structures, and devices having sizes of the order of one to hundreds of nanometers. The basic idea lies in assembling atoms and molecules into complex arrangements that have novel functionality. While mother nature can build all the complex organs of living creatures from protein and other biomolecules, we are unable to build a whole system from the bottom up. Also a precise understanding of the nanoscopic world is essential before we can make full use of nano phenomena. I believe that miniaturized devices and machines play a key role in bridging the gap between the nanoscopic world and our macroscopic world.

This book describes some of the latest developments in micromachining technology based on semiconductor processes for integrated circuits and applied to nanotechnology. The minimum size of transistors has been reduced to almost 100 nanometers thus it becomes possible to fabricate miniaturized tools having sizes ranging from 10 to 100 nanometers. These tools are suitable for measuring, observing, handling and controlling nanoscopic objects, namely DNA molecules, proteins, and semiconductors quantum dots. Of course, this is just the beginning of such applications. I even envision hybrid systems, combining both nanomaterials and micromachined devices, to be developed within the next five to ten years. I hope you share the excitement of observing a new field of research in creation.

I would like to thank Dr. Claus E. Ascheron for his continuous encouragement and Mr. Ryuji Yokokawa for his help in editing chapters.

Tokyo, Japan
22 April, 2003

Hiroyuki Fujita

Contents

List of Contributors

Hiroyuki Fujita
Institute of Industrial Science,
University of Tokyo
4-6-1 Komaba, Meguro-ku,
Tokyo 153-8505, Japan
fujita@iis.u-tokyo.ac.jp

Masao Washizu
Department
of Mechanical Engineering,
The University of Tokyo,
7-3-1 Hongo, Bunkyo-ku,
Tokyo 113-8656, Japan
School of Engineering,
The University of Tokyo
washizu@washizu.t.u-tokyo.ac.jp

Akira Mizuno
Department
of Ecological Engineering,
Toyohashi University of Technology
Tempaku-cho,
Toyohashi 441-8580, Japan
mizuno@eco.tut.ac.jp

Patrick Degenaar
Imperial College London
Exhibition Road,
London SW7 2AZ
p.degenaar@physics.org

Eiichi Tamiya
Japan Advanced Institute
of Science and Technology
1-1 Asahidai, Tatsunokuchi,
Ishikawa 923-1292, Japan
tamiya@jaist.ac.jp

Tatsuo Ushiki
Department of Anatomy
and Histology,
Faculty of Medicine,
Niigata University
Ashahimachi-dori,
Niigata 951-8510, Japan
t-ushiki@med.niigata-u.ac.jp

Hideki Kawakatsu
Institute of Industrial Science,
University of Tokyo
4-6-1 Komaba, Meguro-ku,
Tokyo 153-8505, Japan
kawakatu@iis.u-tokyo.ac.jp

Yasuo Wada
Waseda University,
Wasedatsurumaki-cho, Shinjuku-ku,
Tokyo 162-0041, Japan
wada@waseda.jp

Dai Kobayashi
Institute of Industrial Science,
University of Tokyo
4-6-1 Komaba, Meguro-ku,
Tokyo 153-8505, Japan
dai@iis.u-tokyo.ac.jp

Gen Hashiguchi
Kagawa University,
Takamatsu,
Kagawa 760-8526,
Japan
hasiguti@eng.kagawa-u.ac.jp

Takuji Takahashi
Institute of Industrial Science,
University of Tokyo
4-6-1 Komaba, Meguro-ku,
Tokyo 153-8505, Japan
takuji@iis.u-tokyo.ac.jp

1 Micromachining Tools for Nanosystems

Hiroyuki Fujita

1.1 Introduction

Science and technology have always pushed towards micro miniaturization. In the late 1980s, a single atom was manipulated by scanning tunneling microscope (STM). The STM has a probe as sharp as an atom at the tip. When the distance between the tip and a flat specimen is approximately 1 nm and kept constant by monitoring the tunneling current, it is possible to extract a single atom from the surface of the specimen by applying a voltage pulse and then place it at a desired position. Many experiments were then performed all over the world to write letters [1] and draw figures by arranging atoms. The same apparatus is also useful for observing molecular shapes, manipulating single molecules, and modifying them.

In addition to atomic manipulation or molecular handling, biochemical technology and micro-fabrication technology for semiconductor devices have achieved remarkable progress on the 1–100 nm scales. These technologies have combined and fused together to form a new discipline called nanotechnology. This discipline is concerned with the controlled fabrication of structures with nanometer sizes and the use of novel properties and functionalities arising from nanostructures. Nanotechnology aims to produce novel materials that are built up atom by atom. High-performance devices composed of nanostructures will contribute to our future society, for example, by providing smaller and better information systems and a faster and wider communication network.

Figure 1.1 represents typical sizes of objects and phenomena with both natural and artificial origins. The proteins responsible for biochemical functions in our body have sizes ranging from a few nm to a few tens of nm. A three-base pair of DNA codes a specific kind of amino acid that composes proteins. The length of one code is 0.34 nm. A C_{60} molecule is spherical with diameter a little less than 1 nm. There is a family of such spherical molecules, composed of carbon atoms. Tubular molecules made of carbon, called carbon nanotubes, have diameters of the order of 10 nm and lengths reaching more than 10–100 μm. In the engineering field, the design line width for very large scale integration (VLSI) circuits has decreased to 100 nm. The thickness of gate oxide in a transistor has fallen below 10 nm. The length of a data bit in magnetic recording devices is 20–50 nm. Quantum layers or

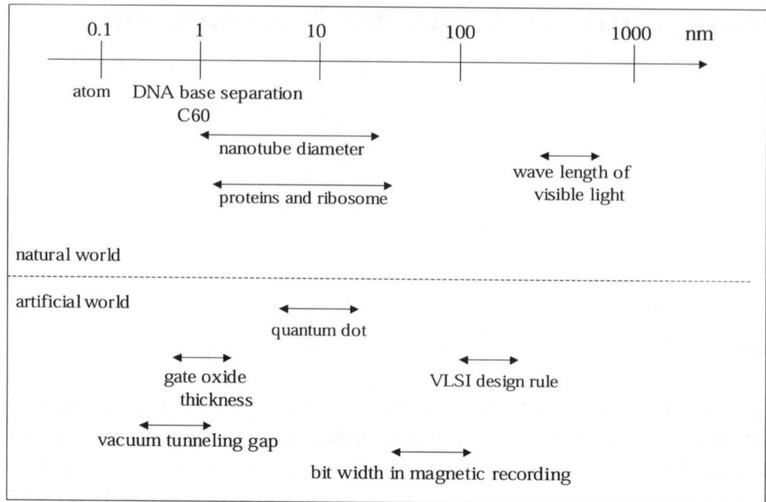

Fig. 1.1. Dimensions of functional structures with both natural and artificial origins

dots in quantum well lasers or single-electron transistors have sizes ranging from 5–20 nm. Nanotechnology tackles structures of these dimensions with a range of different functions.

1.2 Bottom–Up and Top–Down Approaches

Useful materials, devices and systems on the nanometer scale may be obtained by two complementary approaches: building up from smaller sizes or working down from bigger ones. In the so-called bottom–up approach, atoms and molecules are assembled to form higher order structures. A letter of atoms written by STM is a typical example. The optical pressure produced by a highly focused laser beam can manipulate small particles of diameter 100–1 000 nm or even atoms. The technique called laser tweezers may be useful for orientating and arranging molecules in a given order. Even in conventional molecular beam epitaxy (MBE), atomic-level control of thin layers has been achieved. Furthermore, self-organization of atoms and molecules will play a major role in the bottom–up approach, because this does not require individual handling of each element. InAs quantum dots are formed on a GaAs substrate through such self-organizing phenomena. Molecular synthesis depending on chemical reactions may be placed in the same category, because biomolecules are produced in the same manner. The synthesis of a protein from the corresponding gene and the subsequent formation of 3-dimensional structures is the naturally existing paradigm for the bottom–up approach.

In contrast, the ultimate miniaturization of semiconductor fabrication technology and ultra-fine mechanical machining can produce structures in the ten nanometer range. This is called the top–down approach. Bulk materials

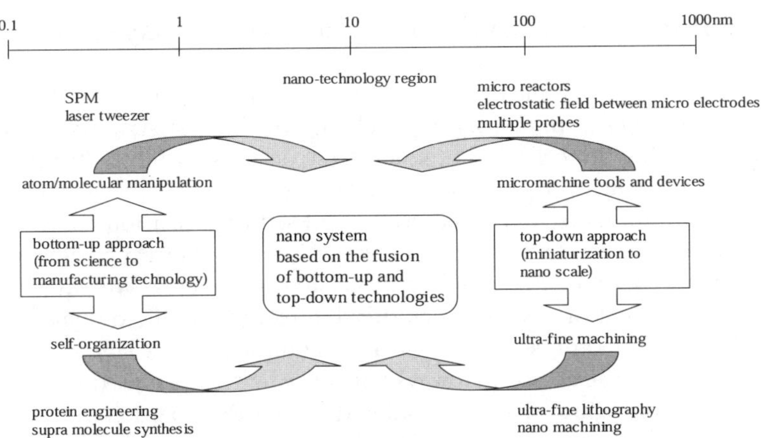

Fig. 1.2. Relationship between the bottom–up and top–down approaches

or thin films are shaped into nanostructures using lithography and etching or very precise cutting tools. Machining precision is better than 1 nm with special tools such as a diamond lathe. The lithography limit has been pushed down to the 10–100 nm range. The extension of semiconductor processes for making 3-dimensional structures has made remarkable progress in the past twenty years. This technology is called micromachining, to be discussed in detail later. Although the first micromachined devices had dimensions of 10–100 μm, nanomachined devices with sizes of 100 nm have been demonstrated recently.

The relation between the bottom–up and top–down approaches is shown schematically in Fig. 1.2. In fact the two approaches are complementary. The bottom–up approach uses atoms and molecules as building blocks for making larger and more complex assemblies. It starts from the size range 0.1–1 nm and extends to the 10–100 nm region, which is the major size range handled in nano-technology, using:

- atom/molecule manipulation by SPM (scanning probe microscope) and laser tweezers,
- self-organization in crystal growth, protein synthesis and chemical synthesis, an approach which is useful for providing nano-materials with novel functionalities originating from quantum phenomena.

The top–down approach comes down from the range 100–1 000 nm to the 10–100 nm region. The smallest features obtained by ultra-fine machining are continually decreasing in size. This is true for both ultra-fine lithography associated with semiconductor processing and mechanical nanomachining. Micromachined tools and devices will play a crucial role in observing, manipulating and synthesizing nanostructures. They are also useful for understanding the interface between the nano-world and our macro-world. This book aims to discuss the effective use of micromachining technology for nanotechnology.

1.3 Combining the Two Approaches to Nanosystems

The human body is a hierarchical system composed of many levels of different dimensions. Starting from the higher level, i.e., the biggest structures, we move from organs, through cells and organelles (structures within a cell, such as mitochondria and nuclei) to functional molecules. The construction of all these levels depends on molecular interactions and self-organization based on genetic information in the DNA. Our technology does not, however, allow us to build a complete system by assembling atoms and molecules. We have developed different fabrication technologies suitable to the specific dimensions we seek to achieve. For example, the heart of the computer is the silicon chip, made by semiconductor processes. However, it must be encapsulated in resin by molding. After packaging, chips are soldered onto a PC board comprising a ceramic material with metal wiring. The external case is made by injection molding and coating and all the sub-units and boards are mechanically assembled.

In the same manner, novel materials obtained by bottom–up nanotechnology should be combined with miniaturized nano-structures and devices. Structures and devices made by top–down nanotechnology serve as the infrastructure for nano-materials (Fig. 1.3a). A simple example is the ion sensitive-field effect transistor (IS-FET). A chemically active material is deposited on the IS-FET gate. When some particular ions, e.g., H^+ for a PH sensor, attach themselves to the layer, the potential of the gate changes and thereby modulates the source–drain current of the FET. More recently, DNA chips have combined both bio- and semiconductor technologies. DNA molecular fragments of various sequences are attached to an array of micro-fabricated

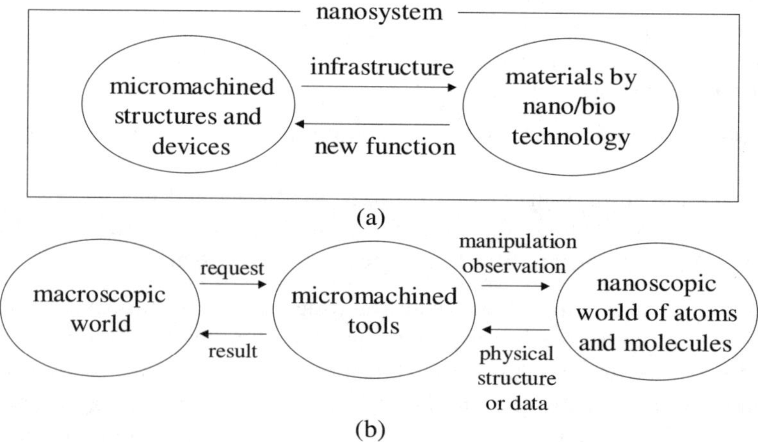

Fig. 1.3. Micromachining technology provides (**a**) infrastructures for nano/bio-materials and (**b**) evaluation tools for observing and manipulating the nanoscopic world

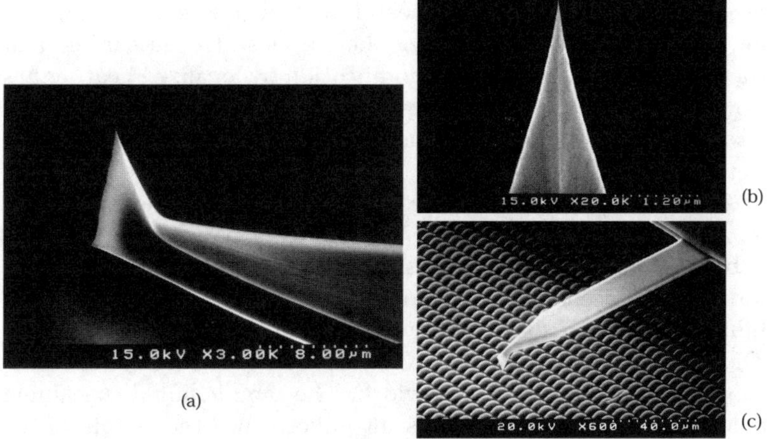

Fig. 1.4. (**a**) Micromachined probe for AFM (atomic force microscope). (**b**) Enlarged probe tip and (**c**) the probe measuring arrays of microlenses [3]

patches. Each patch contains fragments of a particular sequence. When a solution of DNA molecules for examination is introduced onto the chip surface, they are trapped on patches where fragments with complementary sequences are immobilized. After washing the solution away and performing several chemical processing steps, one can observe the places where attached DNA molecules remain by fluorescent imaging.

Bhatia of Harvard University [2] performed co-culture of liver cells (hepatocytes) and other structural cells (fibroblasts) on a chip. The glass chip had micromachined patterns of biomolecules on the surface to mediate cell adhesion of hepatocytes. The remaining uncoated areas attracted fibroblasts. The result showed longer lifetime and better cellular activity. This demonstration has profound consequences for tissue engineering.

Micromachined devices provide not only infrastructures but also tools for examining nano-materials (Fig. 1.3b). SPM probes are made by micromachining technology. The probe for AFM consists of a sharp tip attached to a soft cantilever beam. Figure 1.4 shows scanning electron micrographs of a micromachined AFM probe [3]. The tip is silicon with radius from several nanometers up to 10 nm. The cantilever is made of SiO_2 and has spring constant around 1 N/m. The tip gently touches the surface of a specimen while it is scanned over the surface. Surface topology less than one nanometer can be visualized by observing the cantilever deflection. The tip is used to form nanoscopic pits and recesses by pushing it more strongly against the surface. There are a number of devices in the SPM family. Magnetic force microscopes use the interaction between a magnetic particle on the tip and a nanometric magnetic field distribution on the specimen surface. It is a powerful tool for studying magnetic films for ultrahigh density hard disk drives (HDDs).

A thermal probe microscope has a microheater at the tip. Local thermal conductivity can be characterized by placing the tip close to the surface and measuring the heat dissipation. Surface modification by localized heating has also been demonstrated. SPM is an indispensable tool for nanotechnologies.

Another tool consists of micromachined electrodes in microfluidic channels. An electrostatic field generates a force acting on molecules or small particles. By shaping electrodes and surrounding structures, one can design the electrical field distribution to handle nano-objects. Washizu et al. have manipulated biological cells and molecules by an electric field in a microchannel [4]. In addition, microchannel flows convey objects.

By combining these methods, capillary electrophoresis analysis can be carried out on DNA molecules on a chip (Fig. 1.5) [5]. The glass chip has channels of the order of 10–100 μm in depth and width. The sample liquid containing DNA molecules with fluorescent labels is introduced into the channel from one port. It is transferred to the intersection area by another channel. When the intersection is filled, a buffer solution begins to flow in the orthogonal channel and cuts the sample liquid within the intersection. A small portion of the sample segment then goes into the separation channel. Because the electric field in the channel induces liquid flow through an electro-osmotic effect, DNA molecules are subject to electrostatic force in accordance with their electrical charge. The balance between the electrostatic force and the drag force, which is proportional to molecular length, determines the speed of a DNA molecule. This is the principle of electrophoresis. Both the field and the drag in the microchannel are so large that the separation time is much shorter than in a macroscopic system. The typical separation time is less than one minute. Commercial products have been based on these chips.

The channel could have dimensions in the range 10–100 nm or contain nanometric structures within it. The surface to volume ratio in such a channel is extremely large. Molecules in the liquid therefore interact significantly with the channel walls. Molecular motion is restricted and some kind of ordering

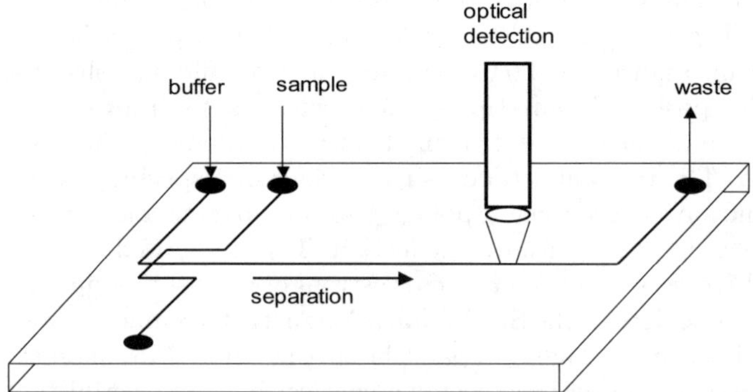

Fig. 1.5. Chip for microcapillary electrophoresis

may occur among molecules. This sort of effect will lead to a new method for controlling chemical reactions, i.e., space-controlled chemistry in nanoreactors.

1.4 Micro- and Nanomachining

Microelectromechanical systems (MEMS) research [6] takes its root in silicon sensor research. Micromachining technologies based on semiconductor processes have been used to make microsensors, their packages and microstructural devices [7]. Sensor research first evolved toward MEMS research at the International Conference on Solid State Sensors and Actuators (Transducers 87), held in Tokyo in June 1987 [8]. At the conference, the presentations on surface micromachining of gears and sliding stages were among the most remarkable. An integrated servo system for mass flow control and some active devices such as an electrostatic actuator were also reported. The first IEEE MEMS Workshop, called the Micro Robots and Teleoperators Workshop, was held in November of the same year [9]. Scientists and engineers have been investigating materials, fabrication processes, device and system design, and applications of MEMS ever since.

This research field is commonly referred to as micromachine in Japan. The term reflects the emphasis on mechanics and a total system approach. A national project launched by the Ministry of International Trade and Industry (MITI) in Japan was also named Micromachine Technologies [10].

Nonetheless, the importance of semiconductor technologies is well appreciated. In particular, MEMS involves two major features:

- many structures can be obtained simultaneously by preassembly and batch processes,
- electronic circuits and sensors can be integrated to obtain smart microsystems.

With these features, high-performance and complex systems which include many actuators with corresponding sensors and controllers can be mass produced in a cost-effective manner.

Table 1.1 summarizes micromachining processes in common use today. The crystallographic dependence of etching speed for single-crystal silicon in such etchants as KOH and TMAH (tetramethyl ammonium hydroxide) is utilized for wet anisotropic etching. The etching speed of the (111) plane is much slower than that of other crystallographic orientations. Well-defined microstructures surrounded by (111) planes can be fabricated.

A new technique called surface micromachining emerged in the 1980s. Thin films of polysilicon and metals are patterned into the shape of a gear, for example. Patterned thin films are released from the substrate by etching away an easily resoluble material placed between the structural film and the

Table 1.1. Micromachining processes, features and applications

Micromachining process (processed materials)	Features	Applications
Crystallographic wet etching (single-crystal silicon)	Precise 3D structures defined by crystal planes	Sensitive membrane for pressure sensors, V-grooves for optical fiber alignment, mirror-flat surfaces
Anisotropic dry etching (silicon)	3D structures of various shapes determined by a mask process	Microactuators generating large forces, freely-shaped microstructures
Surface micromachining (polysilicon thin film, other thin films)	Ultra-fine structures, good compatibility with CMOS circuit process	Integrated sensor array actuators with control circuits (e.g., DMD display)
Hinged 3D structure (polysilicon film)	Microstructures folded up from the substrate to form 3D shapes	Micro-optical devices on a silicon chip
Replica processes: LIGA[a], molding, hexil[b] (metals, polymers, polysilicon, glass)	Many replicas of a 3D master mold are obtained, various materials available	Microfluidic chip by injection molding, glass chip by hot embossing

[a] LIGA is a 3D micromachining method combining X-ray deep lithography, electroforming, and injection molding.
[b] hexil is a 3D micromachining method to obtain 10 – 50 μm thick poly-silicon structures by depositing a CVD poly-silicon layer on a silicon substrate with deep-RIE trenches.

substrate. Researchers were able to make rotating micromotors and linear actuators by this process.

Microstructures fabricated by surface micromachining [6] are planar in nature and have thicknesses of up to 10 μm in most cases. Some applications require thicker structures or 3D-complex structures. Modifications of surface micromachining have been attempted. One technique is to fold up micromachined plates from the substrate to construct a 3D structure [11]. Suzuki et al. [12] released polysilicon plates from the substrate and reconnected them by flexible films made of polyimide. They fabricated a small ant-like structure, brought its legs into resonance by an external vibrational field and drove it. Fukuta et al. [13] fabricated a polysilicon structure by surface micromachining, raised it in the direction normal to the substrate and retained its position by producing Joule heat in the structure and annealing it.

In addition to thin-film 3D processes, deep dry etching of silicon has become very popular over the last five years. Thick microstructures up to

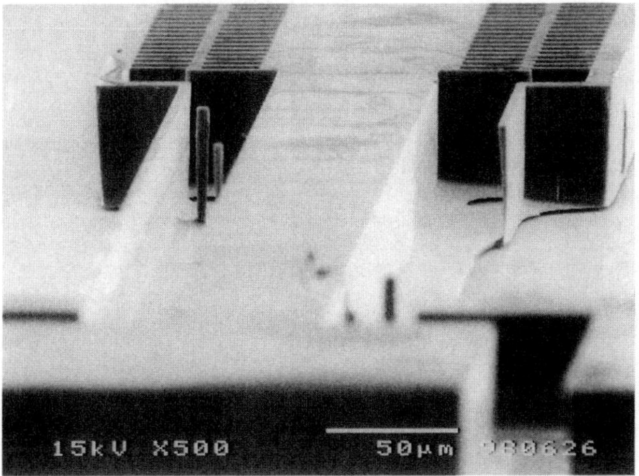

Fig. 1.6. Three-dimensional microstructure fabricated by deep dry etching of silicon. Thickness 50 μm

50–500 μm, with width 1/20 of the thickness, can be fabricated (Fig. 1.6). Electroplating through thick resist patterns creates metallic 3D microstructures. These structures have aspect ratios (the height divided by the width) in the range 20–50. Precision is typically 10–100 nm, but reaches a few nm in certain specific processes.

The variety of materials has also widened. We can now micromachine compound semiconductors (e.g., GaAs), polymers, metals, ceramics and bio-related materials as well as silicon.

1.5 Examples of Micromachined Nanodevices

Some examples of the latest micromachined tools for nanotechnology are described in this section. Examples include microprobes, and in particular arrays of microprobes for ultra-high density data storage, multiple nanoprobes, and microfluidic devices involving bio-functional materials.

1.5.1 Microprobe Arrays for Ultrahigh Density Data Storage

Microprobe-based data storage is a promising alternative to conventional magnetic data storage because it offers potential for considerable storage density improvement, close to or above 1 Tbit/in^2. Data storage devices based on two types of interaction have been widely investigated, i.e., thermomechanical interactions and near-field optical interactions.

Binnig et al. [14] demonstrated storage densities of up to 500 Gbit/in^2 by thermomechanical writing and thermal readout in thin polymer fiims with bit sizes and pitches of 30–40 nm each. This is almost tens times as dense as

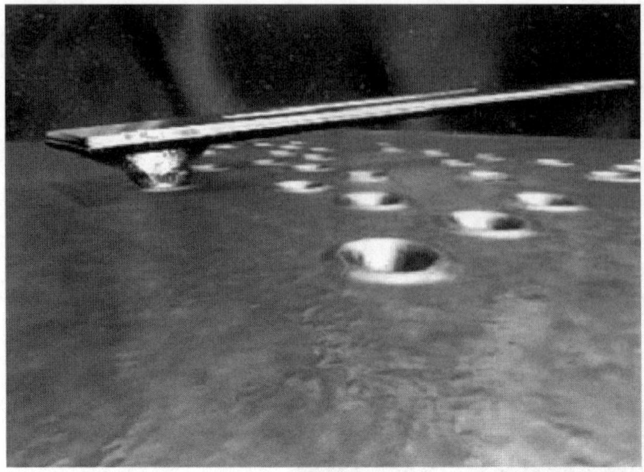

Fig. 1.7. Chip for microcapillary electrophoresis [16]

the current magnetic recording technology. Although the highest data rate achieved by AFM data storage is 6 Mbit/s, it can be improved to a level competing with conventional storage devices by fabricating arrays of many probes using micromachining technology. Vettiger et al. [15] developed and successfully operated a 32 × 32 (1024) two-dimensional cantilever array for parallel AFM imaging and nanodot recording.

Figure 1.7 shows a conceptual drawing [16] of a probe writing a nanodot on a medium. The latter is covered with a 40 nm PMMA film. The writing principle is as follows. The cantilever has a microheater at the end (Fig. 1.8). When it is heated, the tip temperature rises as high as 500°C and melts the thin polymer layer. The heater and tip are so small that there is a very fast thermal response which can achieve highly localized melting. The reading principle is also thermally based. The heat dissipation from the tip varies according to its position: there is less cooling on the flat medium and more cooling in the recess of the nanodot. The temperature can be detected from the heater resistance while the heater current is kept constant. Cantilevers are arranged in an array (Fig. 1.8) and read/write nanodots in parallel as shown in Fig. 1.9. The concept of the data storage device based on an array is shown in Fig. 1.10. The medium is placed on an XYZ stage for alignment and scanning. The array is fixed and connected to multiplexers for driving heaters and reading the resistance.

Another type of microprobe-based data storage is based on near-field optics [17]. Suppose a small aperture of 10–100 nm is illuminated by light. Because the aperture is smaller than the wavelength, the light cannot propagate through it. Instead, a localized electromagnetic field is generated around the aperture. The field decays exponentially according to the distance from the

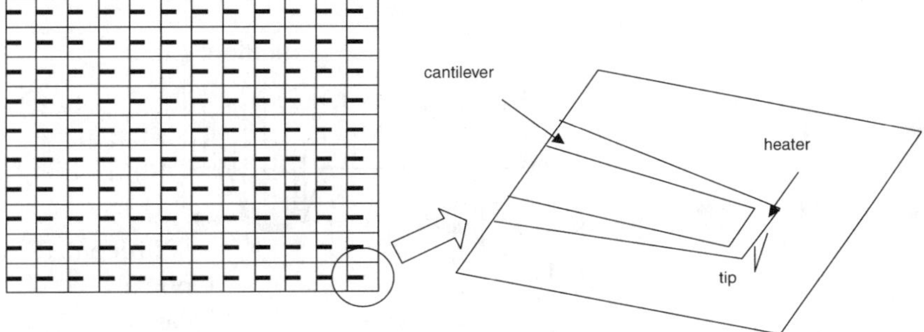

Fig. 1.8. Schematic representation of arrayed probes for parallel data writing and reading

Fig. 1.9. Artist's impression of nano-data storage using arrayed probes [16]

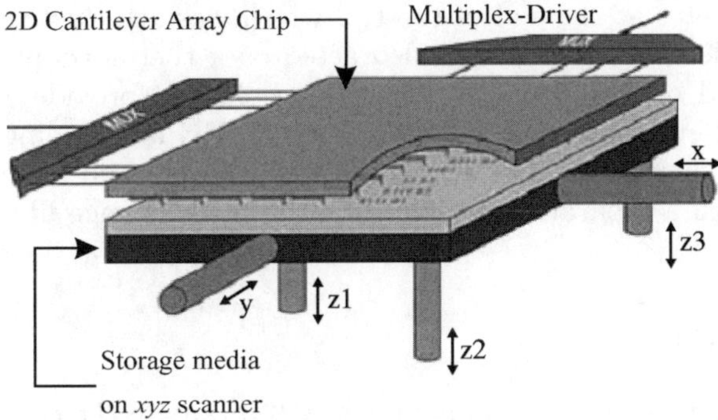

Fig. 1.10. System design for nano-data storage using arrayed probes [16]

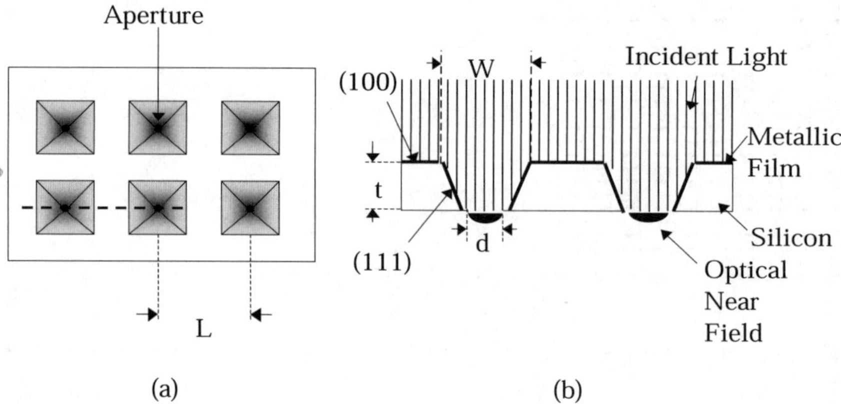

Fig. 1.11. Profile of the probe array with two-dimensional apertures fabricated on a silicon substrate. (**a**) Top view. (**b**) Cross-sectional view

aperture. If a medium is located close enough to the aperture, it is possible to record/read a nanobit of a few tens of nm via the optical near field.

Lee and Ohtsu et al. [18] proposed an array of nano-apertures on the back surface of a slider that flies over a rotating disk in such a way that the slider flies in the magnetic hard disk (Fig. 1.11). Spatially-modulated light is introduced into the array and generates an optical near field on the disk because the distance between the slider and the disk is around 10 nm (Fig. 1.12a). The disk surface is modified according to the field intensity around each aperture, as determined by the modulated light pattern. The scattered light image of the array is used to read the record. Because the axis of the array makes a small angle θ with the flying direction, it is possible to decrease the recording track pitch (Fig. 1.12b). Anisotropic wet etching of silicon was used to fabricate the slider with the array of apertures. The aperture size was 80 nm.

Lee, Ono, Abe and Esashi [19] fabricated a miniature aperture on a silicon cantilever for thermomechanical imaging and data storage. They made a nanoheater 30 nm in diameter at the end of the tip, which was attached to a silicon cantilever. The heater, made of Pt, also acted as a thermocouple because Ni was evaporated on the Pt layer through a hole and provide a bimetal contact. The cantilever had a piezoelectric layer of AlN for bending detection and actuation. They fabricated an array of cantilevers and were able to write nanodots with a separation of 400 nm on a phase change film (GeSbTe).

1.5.2 Multiple Nanoprobes

A four-point probe measurement is the commonest way to evaluate the conductivity of a surface essentially without contact resistance. A current I is

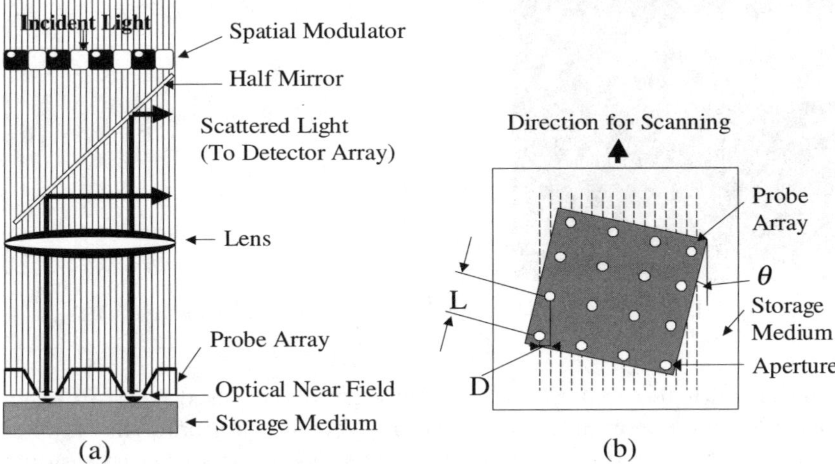

Fig. 1.12. Schematic explanation of storage and readout. (**a**) Storage using spatially-modulated propagating light and probe array. (**b**) Readout by scanning a probe array with tilt angle θ. *Broken lines* represent the trajectory of each aperture

passed through the outer electrodes while the voltage V is measured over the inner probes. The conductivity is extracted from the ratio V/I.

C.L. Petersen [20] and his colleagues at MIC, the Technical University of Denmark [21], have fabricated unique four-point probes (Fig. 1.13). They grew nanosize electrodes extending from the tips of glass electrodes, in order to reduce the electrode spacing to the nanometer level. These electrodes were thin metal-coated glass cantilevers made by micromachining. The mechanical flexibility of such electrodes makes it possible to measure the conductivity of fragile objects, surfaces and thin films on a very small scale. The probes can be moved across a surface to map the conductivity of a region.

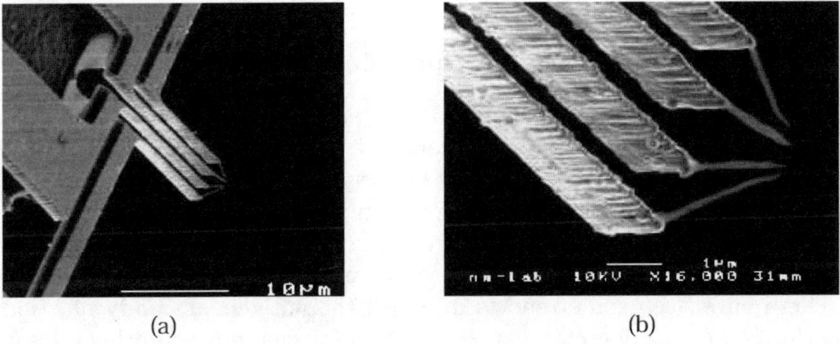

Fig. 1.13. Multiple nanoprobes made by electron beam deposition [21]

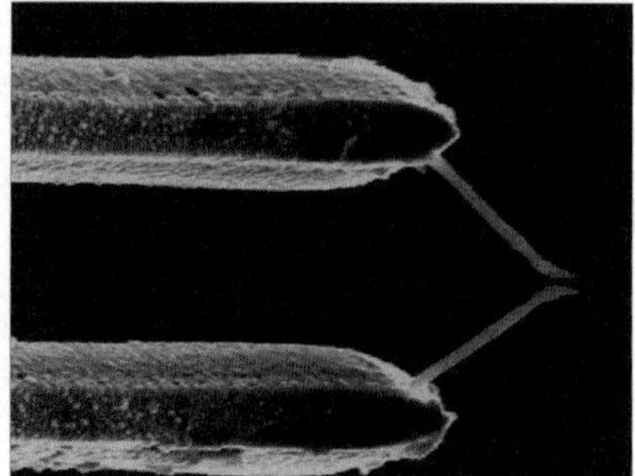

Fig. 1.14. Tips of twin nanoprobes made by electron beam deposition [21]

Together with researchers in Lund University, they used electron beam deposition for the nanotips. By focusing an electron beam onto the end of a microprobe cantilever, the electron beam decomposes carbon-based residues present in the electron microscope. This leads to the growth of a 100 nm thin rod in the direction of the beam. The length can be increased to 5–10 μm. By tilting the probe with respect to the beam, they were able to grow four such nanotips, and thereby reduce the spacing down to 100–300 nm (Fig. 1.13b). In this way they fabricated nanogap 2-probes and 4-probes. The smallness of the electrode spacing allowed them to measure submicron structures on surfaces and films. For the demonstration, they adopted a scanning micro-4-probe setup, this time using only glass cantilevers without nanotips, to map the conductivity of polymer monolayers, and found that ordered domains exhibited a much higher conductivity than disordered regions. The high degree of order in these domains leads to enhanced carrier mobility.

A new multi-function sensing and manipulating nanotool called a nano-hand has been fabricated using nanogap 2-probes (Fig. 1.14). Two narrow fingers with customizable shape forms the tweezer. The latter can be opened and closed using electrostatic voltages. The needles are conducting and can measure conductivity as well as surface topology.

1.5.3 Microfluidic Devices Incorporating Biomaterial

Immunoassay is one of the most important analytical methods for clinical diagnoses and biochemical studies because of its extremely high specificity. However, the conventional immunosorbent assay requires a long time (3 h to 1 d usually), and involves troublesome procedures. It also consumes a considerable amount of expensive reagents. In order to overcome these drawbacks, integration of the immunoassay system into a microchip seems to be effective [22]. There have been some reports in which the antigen–antibody reaction was performed on a microchip [23–28]. Heterogeneous immunosorbent assay is widely used and superior to liquid phase separation because of its easy and clear separation between the free form and the complex. Integration of the heterogeneous immunoassay system, which is based on the same principle as the conventional immunosorbent assay, is expected.

For integrated analytical systems, a detection method with high sensitivity in microspace is indispensable. Kitamori and his colleagues reported a new, highly sensitive immunoassay method on a bulk scale using photothermal spectrometry for detection, in which colloidal gold was used as a labeling material [29–33]. In some photothermal spectrometries, the laser-induced thermal lens microscope (TLM) is especially useful for ultrasensitive determination in microspace because of its high space resolution. They have succeeded in integrating an immunosorbent assay system into a microchip using the TLM as detector. Human secretory immunoglobulin A (s-IgA), which has an important role in local immunity and is known as a stress indicator, and carcinoembryonic antigen (CEA), which is one of the most popular tumor markers, were assayed with the system.

Schematic illustrations of the immunosorbent assay in a microchip using colloidal gold are shown in Fig. 1.15. Polystyrene beads, introduced into a microchannel, were selected for the reaction solid phase. This microchip consisted of a microchannel filled with beads. The analyte (antigen) was adsorbed on the bead surface, and then the antibody conjugated with colloidal gold was fixed on the solid phase by antigen–antibody binding. After the free antigen was washed out, colloidal gold bound to the bead surface via the antigen–antibody complex was detected by TLM.

To keep the polystyrene beads in the microchip while allowing the liquid to flow, a dam structure was fabricated in the microchannel [34,35]. The chip layout is shown in Fig. 1.16. The chip was composed of three quartz glass plates (30 mm × 70 mm), i.e., the cover, middle and bottom plates with thicknesses 170 μm, 100 μm and 1 mm, respectively. Two access holes with diameter 0.5 mm for an inlet and an outlet were mechanically drilled into the cover glass. A deep channel was made in the middle plate with a highly focused and intensified CO_2 laser beam. The middle plate was atttached to the bottom plate by fusion bonding at 1 150°C. A shallow channel in the dam region was fabricated by the fast atom beam fabrication method on the upper side of the laminated plates [35]. The depth of the shallow channel was estimated as 10 μm, so that the reaction solid phase, i.e., polystyrene beads (45 μm in diameter), would be retained in the dam region. Finally, the cover plate was laminated.

The experimental procedures were as follows. After the inner wall of the capillary and the channel was blocked by the casein solution for 1 h, 0.5 microlitres of polystyrene bead suspension was dropped onto the inlet hole of the microchip. The capillary was connected to the inlet hole and phosphate buffer (PB) was pumped in to move the beads to the dam region. Human secretory immunoglobulin A (s-IgA) solution was introduced into the microchannel and liquid flow was stopped to allow adsorption of the antigen onto the surface of the bead. After incubation, the s-IgA solution was washed out with PB, and

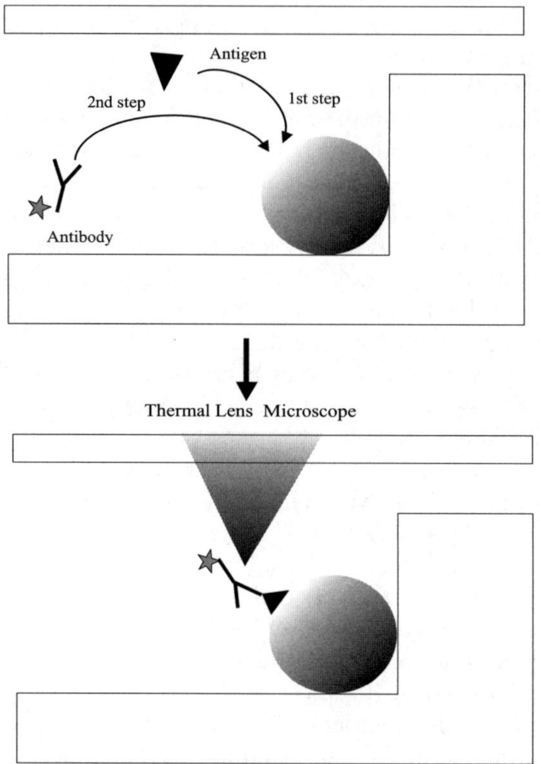

Fig. 1.15. Schematic illustrations of a microchip-based immunosorbent assay

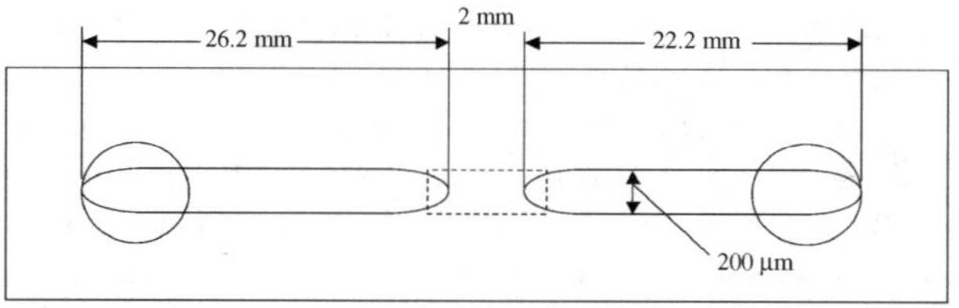

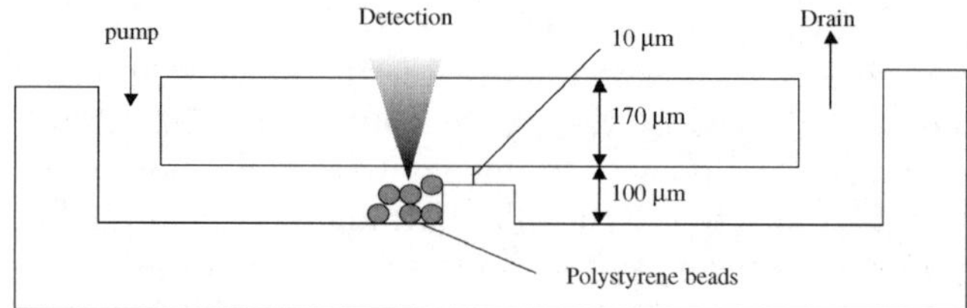

Fig. 1.16. Cancer diagnosis chip containing microbeads coated with antibody [34]

then the residual free surface of the polystyrene beads was blocked by a 10-min flow of casein solution, followed by a 1-min flush of PB. After blocking, antibody–colloidal gold conjugate (AbCG) solution [36,37] was introduced into the microchannel. Then the antigen–antibody reaction was performed. After the reactions, the solution in the microchannel was replaced by PB and the colloidal gold adsorbed on the bead surface was monitored by the TLM.

The duration of the antigen–antibody reaction on a chip was examined. Immunoreaction rates were compared for the integrated immunoassay and conventional assay with a microtiter plate. In the conventional immunoassay using a microtiter plate, the immunoreaction was nearly completed after 15 h. On the other hand, the signal intensity became constant after a 10-min reaction in the microchannel. These results indicate that integration of the immunoassay shortens the reaction time by a factor of ninety. The decrease in the reaction time has two possible reasons based on miniaturization: one is the increase in the ratio of the reactive surface area to the volume of liquid and the other is the decrease in reaction distance to the same order as the diffusion length.

The calibration curve for human s-IgA by microchip immunosorbent assay was measured and compared with the conventional method. The sample containing 1 μg/mL s-IgA showed a clear TLM signal, whereas this sample showed only the same absorbance as the control by means of the conventional immunoassay. Since the concentration of s-IgA in human saliva is normally around 200 μg/mL, this integrated system is expected to be suitable for practical measurements. These results indicate that the integrated immunosorbent assay in the microchip, with a much shorter analysis time, has better sensitivity than the conventional method.

This microchip-based immunoassay system has been applied to determine a tumor marker, carcinoembryonic antigen (CEA) in serum. Because three antigen–antibody reaction steps were involved in the assay, the total required time was almost 50 h for complete reaction using the conventional method. On the other hand, since the time neccesary for each reaction is only 10 min by microchip assay, the total assay time was reduced to about 40 min. Moreover, the determination limit was about 0.1 ng/ml, which is much lower than the cutoff value for colon cancer (roughly 5 ng/ml). This system is thought to be useful for clinical diagnosis. The integrated immunosorbent assay system will be put into practical use as part of an automated system in the future, because of the short analysis time, easy procedures, and low cost.

1.6 Organization of the Book

Following the basic introduction given in this chapter, a detailed description of specific research topics is given in each chapter.

Chapters 2 and 3 are concerned with the single-molecular handling of DNA by physical means. In one case, molecular handling mainly operates through the electrostatic force, while the modification, e.g., cutting, of molecules is achieved by mechanical force using a microprobe or chemical reaction of enzymes. In the other case, a focused laser spot produces a handling force for molecules. These techniques will allow us to read the genetic information in DNA at the specific location defined by the physical distance from one end of the molecule.

Chapters 4 and 5 deal with measurement and visualization of biological samples by means of scanning probe microscopes. Chapter 4 describes observation by scanning near-field optical microscopy (SNOM) combined with atomic force microscopy (AFM). Target cells are grown on micromachined patterns or in micromachined channels. The motion of living cells is observed in real time by AFM in Chap. 5.

Chapters 6 to 8 are devoted to tools for investigating the physical world. They describe the engineering development of instrumentation for nanometric measurement and explain the fabrication processes for nanostructures by semiconductor technology. Application targets include quantum nanodots, nanowires and molecules that could be used in molecular electronics.

It is hoped that readers will themselves come into contact with the emerging technology of micro- and nanofabrication and its expanding applications to nanotechnology. The greatest reward for compiling this book would be to know that it had led others to make breakthroughs in their own research.

References

1. D.M. Eigler, E.K. Schweizer: *Positioning Single Atoms with a Scanning Tunneling Microscope Tip*, Nature **34**, 6266 (1990) pp. 524–6.
2. Sangeeta Bhatia: *Microfabrication in Tissue Engineering and Bioartificial Organs* (Kluwer Academic, 1999)
3. `www.olympus.co.jp/Profile/Recruit/products/microcomp/` (2001)
4. M. Washizu, O. Kurosawa, I. Arai, S. Suzuki, N. Shimamoto: *Application of Electrostatic Stretching-and-Positioning of DNA*, IEEE Trans. IA **31**, 3 (1995) pp. 447–456
5. D.J. Harrison, K. Fluri, K. Seiler, Z. Fan, C.S. Effenhauser, A. Manz: *Micromachining a Miniaturized Capillary Electrophoresis-Based Chemical Analysis System on a Chip*, Science **261**, 895–898 (1993)
6. R.T. Howe, R.S. Muller, K.J. Gabriel, W.S.N. Trimmer: *Silicon Micromechanics: Sensors and Actuators on a Chip*, IEEE Spectrum (June 1991) p. 29
7. K. Petersen: *Silicon as a Mechanical Material*, Proc. IEEE **70**, 420 (1982)
8. Proc. of 4th Intl. Conf. on Solid-State Sensors and Actuators, Tokyo, Japan, June 2–5, 1987
9. Proc. of IEEE Micro Robots and Teleoperators Workshop, Hyannis, MA, Nov. 9–11, 1987
10. Micromachines, published quarterly by Micromachine Center (MMC) in Japan. MMC (website) manages the Micromachine Technology Project in Japan. `http://fsic.mmc.or.jp`
11. D.S. Gunawan, L-Y. Lin, K.S.J. Pister: *Micromachined Corner Cube Reflectors as a Communication Link*, Sensors and Actuators A **46–47**, 580–583 (1995)

12. K. Suzuki, I. Shimoyama, H. Miura: *Insect-Model Based Microrobot with Elastic Hinges*, IEEE/ASME J. of Microelectromechanical Syst. **3**, 4–9 (1994)
13. Y. Fukuta, T. Akiyama, H. Fujita: *A Reshaping Technology for Three-Dimensional Polysilicon Structures*, Trans. IEE of Japan **117** E, 20–25 (1997)
14. G. Bennig et al.: Appl. Phys. Lett. **74**, 1329 (1999)
15. P. Vettiger et al.: IBM J. Res. Develop. **44**, 323 (2000)
16. `www.zurich.ibm.com/st/mems/`
17. E. Betzig, J.K. Trautman, T.D. Harris, J.S. Weiner, R.L. Kostelak: *Breaking the Diffraction Barrier: Optical Microscopy on a Nanometer Scale*, Science **251**, 1468–1470 (1991)
18. M.B. Lee, T. Nakano, T. Yatsui, M. Kourogi, K. Tsutsui, N. Atoda, M. Ohtsu: *Fabrication of Si Planar Apertured Array for High Speed Near-Field Optical Storage and Readout*, Technical Digest of the Pacific Rim Conference on Lasers and Electro-Optics, Makuhari, Japan, No. WL2, 91–92 (1997)
19. D.W. Lee, T. Ono, T. Abe, M. Esashi: *Fabrication of Microprobe Array with Sub-100 nm Nano-Heater for Nanometric Thermal Imaging and Data Storage*, IEEE International Conference on Micro Electro Mechanical Systems, January 21–25, 2001, Interlaken, Switzerland (2001) pp. 204–207
20. C.L. Petersen: *Microscopic Four-Point Probes*, Ph.D. Thesis, Mikroelektronik Centret, DTU, Lyngby, Denmark (1999)
21. `www.mic.dtu.dk/research/nanotech/nanotech.htm`
22. K. Sato, M. Tokeshi, H. Kimura, T. Kitamori: *Biomimetic Chip: Development of Ultrafast Cancer Diagnosis Chip*, Tech. Digest of IEEJ 17th Sensor Symposium, May 30–31, 2000, Kawasaki, Japan (2000) pp. 95–100
23. L.B. Koutny, D. Schmalzing, T.A. Taylor, M. Fuchs: Anal. Chem. **68**, 18–22 (1996)
24. F. von Heeren, E. Verpoorte, A. Manz, W. Thormann: Anal. Chem. **68**, 2044–2053 (1996)
25. N. Chiem, D.J. Harrison: Anal. Chem. **69**, 373–378 (1997)
26. C.L. Colyer, T. Tang, N. Chiem, D.J. Harrison: Electrophoresis **18**, 1733–1741 (1997)
27. N. Chiem, D.J. Harrison: Electrophoresis **19**, 3040–3044
28. N. Chiem, D.J. Harrison: Clin. Chem. **44**, 591–598 (1998)
29. K. Sato, M. Tokeshi, T. Odake, H. Kimura, T. Ooi, M. Nakao, T. Kitamori: Anal. Chem. **72**, 1144–1147 (2000)
30. S. Matsuzawa, H. Kimura, C.Y. Tu, T. Kitamori, T. Sawada: J. Immunol. Methods **161**, 59–65 (1993)
31. C.Y. Tu, T. Kitamori, T. Sawada, H. Kimura, S. Matsuzawa: Anal. Chem. **65**, 3631–3635 (1993)
32. H. Kimura, S. Matsuzawa, C.Y. Tu, T. Kitamori, T. Sawada: Anal. Chem. **68**, 3063–3067 (1996)
33. H. Kimura, M. Mukaida, T. Kitamori, T. Sawada: Anal. Sci. **13**, 729–734 (1997)
34. K. Sato, M. Tokeshi, T. Kitamori, T. Sawada: Anal. Sci. **15**, 641–645 (1999)
35. Y.Toma, M. Hatakeyama, K. Ichiki, H. Huang, K. Yamauchi, K. Watanabe, T. Kato: Jpn. J. Appl. Phys. **36**, 7655–7659 (1997)
36. G. Frens: Nature Phys. Sci. **241**, 20–22 (1973)
37. U.R. Chakraborty, N. Black, H.G. Brooks Jr., C. Campbell, K. Glick, F. Harmon, L. Hollenbeck, S. Levison, D. Mochnal, P. O'Leary, E.D. Puzio, W. Tien, A. Venturini: Ann. Biol. Clin. (Paris) **48**, 403–408 (1990)

2 Microsystems for Single-Molecule Handling and Modification

Masao Washizu

In the present chapter, we investigate a novel method for molecular manipulation and modification at the single-molecule level using microtechnologies, taking DNA as our target. A high-intensity high-frequency electrostatic field ($\geq 10^6$ V/m, ≈ 1 MHz) created in a microfabricated structure is used to immobilize DNA at a predetermined position on a solid surface with fully stretched conformation. A laser, a mechanical stylus, or an enzyme-immobilized probe is used to cut the stretched DNA at the target position. The cutting position of a restriction-enzyme probe is shown to agree with the restriction map, providing proof that the specificity of the enzyme is preserved even when both the enzyme and the target molecule are immobilized.

Conventional chemistry treats the ensemble average of molecules in solutions: a large number of molecules, measured with the unit [mol], are dissolved in a solvent, and reactions occur by chance through stochastic collisions between molecules. Little attention is paid to which molecule reacts with which, or attacks which part of some other molecule.

The problem associated with the inability to discern a molecule in a solution, or a position on a molecule, becomes most conspicuous when handling DNA. In DNA, genetic information is recorded as a linear sequence of bases. The 'position' of the base thus has an essential meaning. However, when it is treated in water solution, as is always done in conventional biochemistry, we are unable to specify a particular molecule in the solution, nor a particular position on a molecule. As a result, the following problems can arise:

- When determining a long DNA sequence, it is first cut into smaller fragments which are compatible with electrophoretic assays (typically less than 1 000 base pairs). But diffusion of fragments takes place as soon as it is cut, and the location of a fragment in the original DNA strand becomes untraceable. Hence the total sequence must be inferred from the patchwork of sequences of fragments.
- Restriction enzymes can cut DNA at a particular sequence of several base pairs, but when there are more than two restriction sites, which site will be cut is a matter of chance.
- Ligation can take place among any cohesive ends.
- The position at which a foreign gene is inserted, say in a chromosome, is unpredictable.

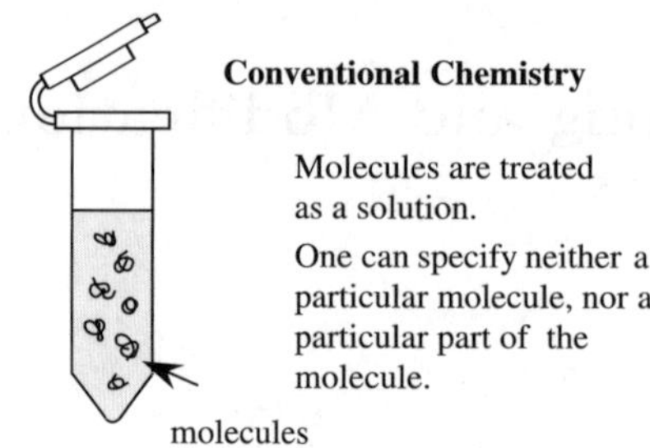

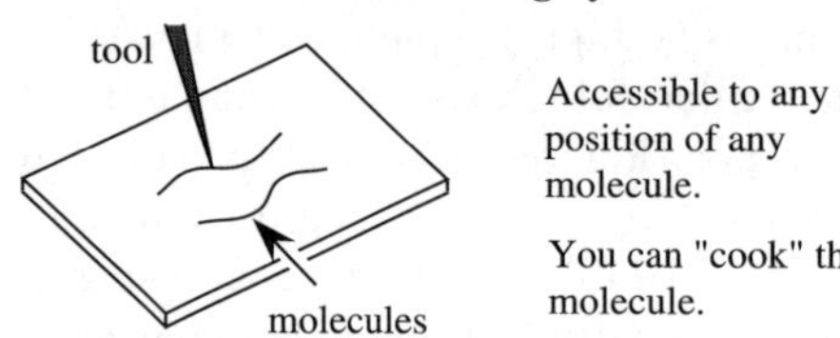

Fig. 2.1. Molecular surgery

- The distance between two genes in a DNA strand, or the distance to the regulation sequence, has an important meaning in gene expression, recombination, etc. However, we have no way of directly measuring and controlling such distances.

These difficulties can be overcome if we have a method for controlling the position and conformation of individual molecules. Once this is done, we have access to arbitrary positions on the molecule, making it possible to carry out pinpoint modifications at a given target position. This kind of molecular modification with spatial resolution, referred to as molecular surgery, is schematized in Fig. 2.1.

Recent advances in micro- and nanotechnology, including

- micromachining,
- electrostatic manipulation,
- laser tweezers,
- micropositioning,
- scanning probe microscopy,
- molecular binding and patterning techniques,
- visualization techniques such as fluorescent microscopy in combination with new fluorescent dyes,

have made such physical manipulation on the single-molecule level a reality.

Basically, molecular surgery can be applied to any molecule insofar as we have a small enough tool. However, DNA can be regarded as the most important object, because DNA is the ‘blueprint’ for biological systems and can be easily amplified, so we need to modify only one molecule in principle. DNA has a dimension of 0.34 μm/kb, where 1 kb = 1 000 DNA bases, which is within the reach of recent microfabrication techniques.

We have developed methods for molecular surgery of DNA, and experimentally demonstrated the cutting of DNA at arbitrary locations, using such tools as laser, atomic force microscopy (AFM) stylus, or enzyme-immobilized particles. Experiments show that the key issue is immobilizing the DNA in such a way that the tool can work effectively, and microsystems can be used as a powerful tool for the purpose. The principles and the method are reported here, together with experimental results.

2.1 Stretch-and-Positioning of DNA

The first step in molecular surgery is to immobilize the molecule. DNA is a long, thin string-like molecule, and adopts a randomly coiled conformation in water due to thermal agitation. Therefore it must first be stretched out and immobilized at a predetermined position on a solid surface in order to allow easy access to the target position on the molecule. For this purpose, we use the kinetic effects [1,3] of an a.c. electrostatic field created in a microfabricated electrode system.

Figure 2.2 is a schematic depiction of the process which we call electrostatic stretch-and-positioning of DNA [4,5]. A pair of electrodes with a typical gap of 100 μm is fabricated on a glass substrate by a photolithographic process. DNA solution is fed in, with concentration in the range 0.05 pg/ml to 10 μg/ml depending upon the required density of immobilized DNA, and then covered by a cover slip. The water film thickness should be 10–20 μm to suppress electrohydrodynamic flow of the solution induced by voltage application. At this stage the DNA adopts a random shape (Fig. 2.2a).

When an electrostatic field is applied, DNA polarizes by the process known as counter-ion polarization [6]: phosphate groups forming the backbone of DNA are negatively charged, and positive ions are attracted to form a counter-ion cloud. When the field is applied, the cloud moves along the DNA strand, resulting in excess positive charge near the molecular end downstream of the field line, and exposed backbone with negative charge on the other end.

Kinetic effects are produced by this induced polarization. The Coulombic force exerted on the positive and negative charges pulls the strand into a straight shape (Fig. 2.2b). The two pulling forces are not always equal: if one of the legs is near the electrode edge where the electrostatic field is more intense, the force on this side will be stronger. As a result, the DNA moves towards the electrodes (a phenomenon called dielectrophoresis [1], in which a polarizable object is driven by a field gradient), until the molecular terminus comes into contact with the electrode edge (Fig. 2.2c).

When an active metal such as aluminum is used as the electrode material, the contact point is permanently anchored. The anchoring is strong enough

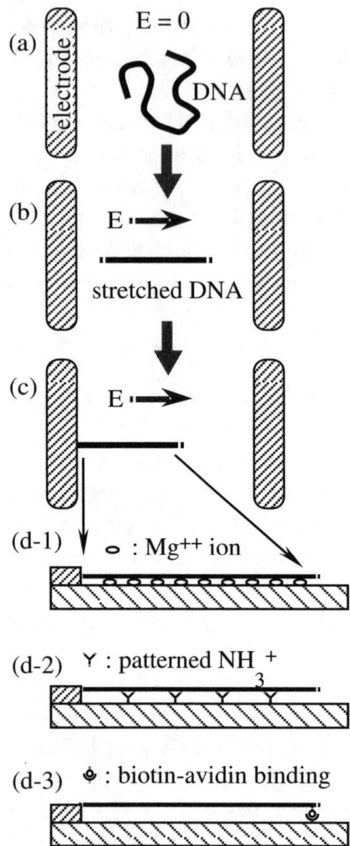

Fig. 2.2. Electrostatic stretch-and-positioning of DNA. (**a**) DNA in water takes randomly coiled conformation due to thermal agitation. (**b**) Stretching of DNA occurs under an electrostatic field of 10^6 V/m, 1 MHz. (**c**) DNA is positioned onto the electrode edge by dielectrophoresis. The contact point is permanently anchored. (**d**) Immobilization of DNA (Side View); (**d-1**) total immobilization, (**d-2**) multi-point immobilization, (**d-3**) immobilization only at the terminus

to ensure that, when a hydrodynamic drag is exerted, the strand will break at the middle rather than coming off the electrode. The anchoring mechanism is not clear, however. Our experimental observation that anchoring is weaker with less active metals such as Au or Pt, or with an oxidized Al surface, suggests the involvement of electrochemical reactions covalently binding the end of the DNA molecule to the metallic surface.

The field intensity required to stretch-and-position DNA is about 1 MV/m at 1 MHz. A high frequency must be used to avoid chemical reactions at the liquid/metal interface. One advantage in using microelectrodes is that the voltage can be low. If the electrode gap were 1 cm, a 10 kV 1 MHz power supply would have been required. Another advantage is related to the scaling law of thermal conduction. The high-field region in a microfabricated electrode is confined to a small volume, so that Joule heat under the high intensity field is efficiently removed, and the temperature rise is kept low. Even so, the maximum allowed salt concentration is several mM. We usually use deionized water with resistivity above 100 $k\Omega\,cm$, and the estimated temperature rise is a few degrees centigrade. When biochemical processes require a buffer solution, say for enzymatic reactions, the solution must be replaced after removing the field.

When the field is removed, the stretch-and-positioned DNA shrinks back into a random-coiled conformation with one end still anchored on the electrode. Therefore, in order to maintain a stretched conformation, the DNA must be immobilized on the substrate while the voltage is on. There are several ways to do this, as shown in Fig. 2.2d.

If the DNA is to be anchored along its entire contour on a glass substrate, the simplest way is to add a small amount of positive divalent ions such as Mg^{++}. The ions act as a binding agent between the negatively charged DNA backbone and the negatively charged glass surface. The anchoring strength can be controlled by the amount of ion added. 0.1 mM Mg^{++}, whose conductivity does not hamper electrostatic stretching, is found to be adequate for stable immobilization. To prevent the DNA from sticking in unwanted locations, the ion is added once the electrostatic stretch-and-positioning process has been completed.

Multi-point anchoring of DNA at certain intervals can be achieved by patterning the substrate with DNA-binding groups. An example is a pattern of positively charged amino groups, which may be fabricated by lifting off the amino silane coupler.

If DNA is to be anchored at both ends leaving the middle part free, avidin–biotin binding or metallic patterning can be used. The former is based on the strong binding between avidin and biotin. A biochemical technique exists to incorporate biotin molecules at the molecular termini of DNA. Several bases from the 3′ end of DNA are removed enzymatically and then repaired with biotin-labeled bases using DNA polymerase. The glass substrate, carrying a pair of microfabricated electrodes, is first coated with avidin, and then the DNA is stretch-and-positioned. Biotin at the stretched end binds with avidin underneath and the DNA is immobilized [7]. Immobilization occurs regardless of the length of DNA. When the DNA size is uniform and known in advance, use of metallic patterning is found to be preferable. This will be detailed later.

There are also possibilities for using many other molecular bindings, including the one between Au and SH, or those due to photoreactive groups. Our experience is that the former reaction is not as quick as the one using avidin–biotin, or anchoring onto an Al surface. The latter, for example, using photoactivation of N_3, is effective in high concentration solutions, or in making a covalent bond between already touching molecules. However, when photoreactive groups are first immobilized onto a solid surface and try to catch DNA in a dilute solution, the collision must take place within the lifetime of the radicals, and the chances are rather small. In both methods, instantaneous anchoring is seldom observed under the microscope.

2.2 Molecular Surgery of DNA

2.2.1 Laser Surgery

Figure 2.3 shows stretch-and-positioned DNA cut by irradiation from a focused ultraviolet (UV) laser. Here, λ-DNA (48 kb) is stained with the fluorescent probe DAPI and the observation is made under a fluorescence microscope. In the photo, many DNA molecules are visible, each stretched and anchored on the electrode edge, as depicted schematically in the lower part of the figure. The spacing between the aligned DNA molecules is estimated to be a few tenths of a micrometer, beyond the resolution of the photo, so individual DNA molecules cannot be clearly seen. However, the total fluorescence appears as a white belt. The width of the belt is measured to be 16 μm, in good agreement with the fully stretched length of λ-DNA calculated from the structural constant, 0.34 nm/base $\times$ 48 kb = 16.3 μm. One end of the DNA is anchored on the electrode edge, while the other end is left free. The electrostatic field is continuously applied during the process.

It is observed experimentally that DNA is instantaneously cut at the position irradiated by a pulsed N_2 laser (337 nm UV, 100 μJ/shot), focused down to < 1 μm in diameter. The portion anchored on the electrode edge remains, while the other portion diffuses into the solution and disappears, so that this part is lost in the photo.

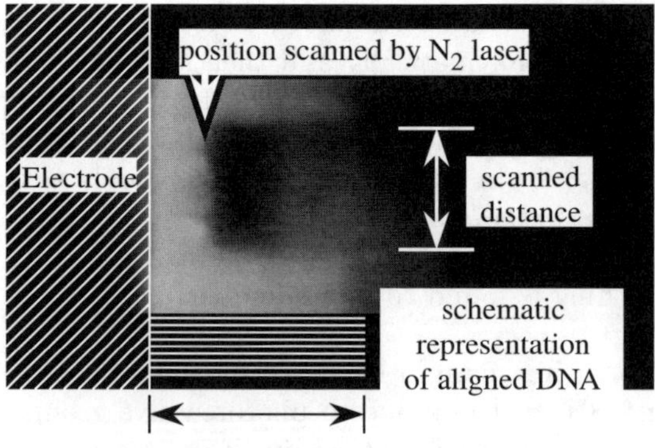

Fig. 2.3. Laser surgery of stretched DNA

Such DNA dissection at an arbitrary position might have applications in cutting out or destroying a particular gene. The method is simple in that no anchoring is required, but it has an essential limitation with regard to cutting accuracy. The theoretical minimum spot size of focused light is approximately equal to its wavelength. For the UV laser, the wavelength is 337 nm, and DNA has about one thousand bases in this length. Such low resolution not only affects precision in the cutting location, but also results in unpredictable molecular damage of the bases. It has been shown that the damage can be removed enzymatically with the use of exonuclease [4]. However, the loss of 1 000 bases can still be a problem, if we compare with the number of bases that can be analyzed in a batch by electrophoresis.

2.2.2 Mechanical Surgery with an AFM Tip

High-resolution dissection is achievable using a tool with a sharp tip, such as an AFM stylus. DNA can be dissected by pressing the stylus against the stretch-and-positioned DNA and moving perpendicularly to it, just as one would cut spaghetti on a plate with a knife. Such mechanical dissection may have an application in 'ordered' sequencing, where DNA is sliced into small pieces from one end, and then each fragment is analyzed to obtain the total sequence of the original DNA. This procedure avoids the tedious patchworking required in conventional shot-gun sequencing.

A technical problem here is how to recover the dissected fragments efficiently. If the fragments are left free, they diffuse into the solution and will be lost. For high-yield fragment recovery, we have developed a method using a DNA carrier layer, released from the substrate by sacrificial etching [8]. The device, depicted schematically in Fig. 2.4, consists of a glass substrate onto which a sacrificial layer, a DNA carrier layer, and a pair of Al electrodes are deposited. DNA solution is fed onto the device and covered with a cover slip. The covering is necessary in order to prevent hydrodynamic flow during voltage application, otherwise the DNA will not be positioned neatly. Using the electrostatic field, the DNA is stretch-and-positioned and one leg of the DNA is anchored onto the Al electrodes. Then the voltage is turned off and the cover slip is slowly removed by sliding it parallel to the aligned DNA. The water gradually evaporates and the DNA is immobilized on the carrier plate during the drying process. The immobilization here is the same as in molecular combing [9], except that one end of the DNA is anchored on the electrode edge, causing the DNA to be immobilized in a square arrangement. Then the DNA is cut, together with the underlying carrier layer. Finally, the sacrificial layer is dissolved with a solvent and a strip of the carrier layer with DNA fragments on it is recovered on a membrane filter. This process recovers the DNA fragments without loss.

The method is demonstrated experimentally using a λ-DNA specimen (48 kb = 16 μm length), a spin-coated gelatin layer with thickness 0.1 μm as the carrier layer, and 1 μm of photo-resist as the sacrificial layer. The gelatin

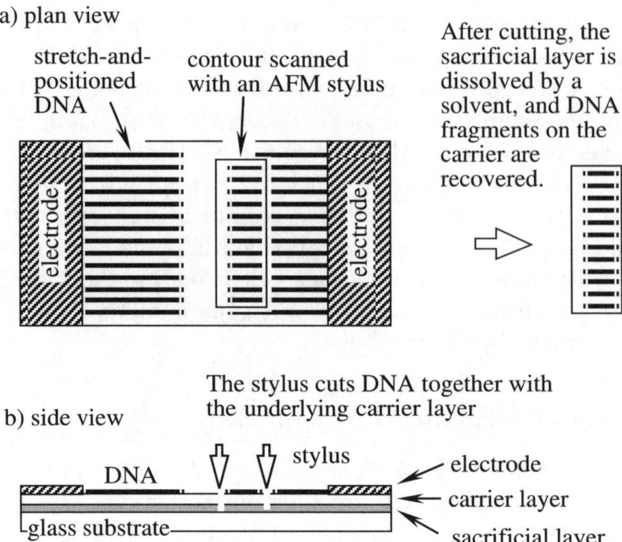

Fig. 2.4. Dissection and recovery of a target position on DNA using a carrier and a sacrificial layer

layer is soft enough to ensure that an AFM stylus can penetrate it. Figure 2.5a shows a photo of the carrier layer. The contour of the approximately 10 μm × 40 μm area is cut using an AFM stylus with half-cone angle 18°. The position of the cut is designed to include 3 μm from one end of the λ-DNA. Then the sacrificial layer under the dissected area is dissolved by pouring ethanol onto it, and the strip is recovered on a filter (Fig. 2.5b) and put into a test tube.

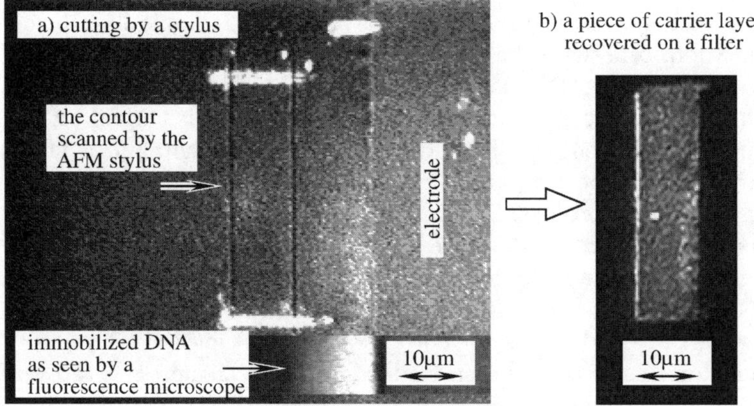

Fig. 2.5. Mechanical cutting of DNA on the carrier layer

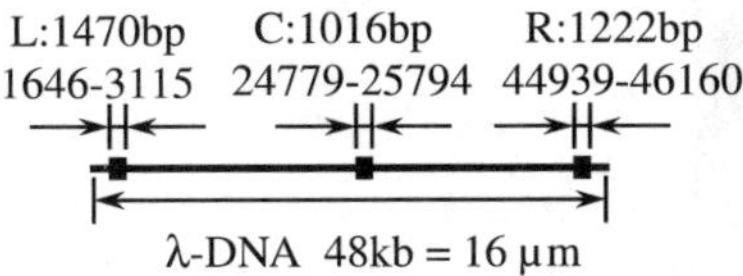

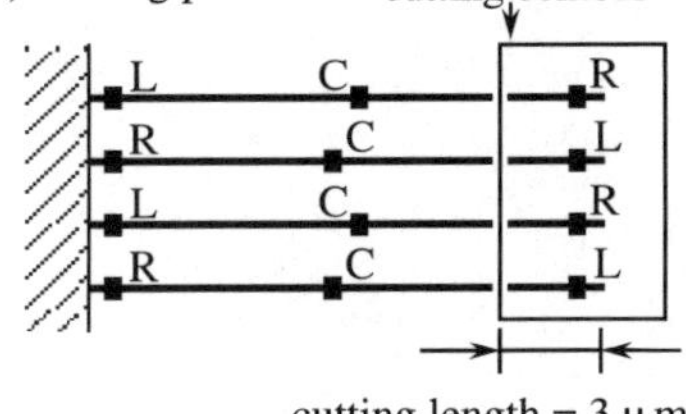

Fig. 2.6. Cutting position and primer design

By raising the temperature, the gelatin layer is melted to obtain the DNA fragments in a water solution.

The density of aligned DNA typically obtained with the electrostatic method is several molecules per 1 μm of electrode contour. It seems that the repulsion between dipole moments induced on DNA molecules prevents them from getting too close to each other. Hence, the number of DNA fragments practically obtainable with the above method will be several hundred, which is adequate for a stable PCR amplification.

To prove that the desired position in the sample DNA is dissected and recovered, three primers are prepared, as depicted in Fig. 2.6a. They correspond respectively to the λ-DNA sequences near the left end (denoted L), near the right end (R), and approximately at the center (C), all about 1 kb in length. Because DNA is electrically symmetrical, electrostatic alignment yields a mixture of one orientation and the other, as shown in Fig. 2.6b. If the aligned DNA is cut 3 μm from the end and successfully recovered, the PCR product should contain the sequences L and R, but not C (Fig. 2.6b). On the other hand, if the sequence C is detected by PCR, it is an indication that unwanted DNA molecules are coming in.

Figure 2.7 shows electrophoresis of the PCR result. Lanes C1 and C2 are the positive control, starting from 20 and 200 λ-DNA molecules, respectively, where the three bands from top to bottom are L (1470 bp), R (1222 bp), and C (1016 bp). Five carrier strips are mechanically dissected and recovered by the method of Fig. 2.4, whose PCRs are shown in lanes 1 through 5. In lanes 1, 2, 3 and 5, only the bands corresponding to L and R are seen, proving successful dissection and recovery. The case of lane 4 is a failure, where a small amount of impurity, i.e., an unwanted portion of DNA, is detected as band C. Comparison with the brightness of bands in S2 reveals that more than 200 DNA fragments are recovered by each mechanical dissection.

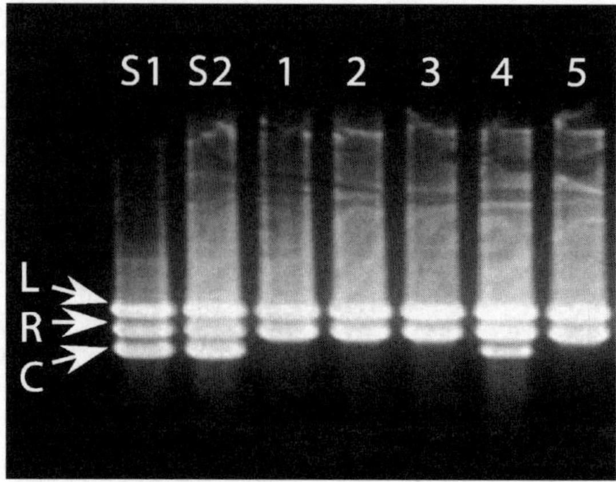

Fig. 2.7. PCR result for DNA fragments obtained by mechanical dissection. S1: λ-DNA 20 molecules. S2: λ-DNA 200 molecules. 1–5: fragments obtained by mechanical dissection

In order to apply the sequencing method, dissected DNA fragments must be amplified and cloned. In the experiment of Fig. 2.7, we were able to perform PCR because we knew the sequence to design the primers. In order to analyze unknown DNA fragments, we have to add a short oligonucleotide with known sequence (an adapter) to both ends of the fragment to which primers are to be bound. Our experiments show that such a biochemical process has a yield of only a few percent. The result illustrated in Fig. 2.7 shows that the number of DNA fragments obtained is adequate for this process.

The fact that electrostatically positioned DNA is a mixture of two orientations yields ambiguity in the original position of the fragment (i.e., the left or right end). A method has been developed to avoid this ambiguity [5]. After anchoring, a restriction enzyme is added which cuts near one end. DNA with the orientation that has the restriction site near the anchoring end is released, while DNA with the other orientation remains, losing only a small fraction at its free end. The method is of course only applicable when a suitable restriction enzyme exists.

2.2.3 Molecular Surgery with an Enzyme-Labeled Probe

The reader is referred to [10] for further details concerning the discussion in this section.

The Basic Principle. An enzyme-immobilized probe, whether it be a sharp tip or a small particle, can be used as a tool for molecular surgery. By immobilizing enzymes on a solid phase and manipulating them by physical means, one can specify the location at which the enzyme works. Many applications can be considered for molecular surgery with enzyme-immobilized probes, including molecular modification such as DNA methylation (addition of methyl groups on bases, known to be one of the regulation mechanisms for gene expression), mutation, insertion, and so on. However, modification at the single-molecule level, even if successful, is not easy to detect. We therefore investigated DNA cutting, because the cutting event can be visualized on a real-time basis. Throughout the investigation, we tried to establish a methodology and also to clarify the conditions for successful interaction between the immobilized enzyme and the immobilized target molecule (what biochemists call the 'substrate').

One of the advantages of DNA cutting with enzyme-immobilized probes is that it yields a chemically defined molecular end. This contrasts with mechanical or laser cutting, where the molecular structure of the cut end is unpredictable. Another advantage is its potential for 1-base resolution. A base is one of A, T, G, or C, so a 4-base cutter restriction enzyme (recognizing 4 base sequences) has a cutting site in every 4^4 bases $\times$ 0.34 nm/base = 87 nm on average, and a 6-base cutter in every 4^6 bases = 1.4 µm. One can thus use the physical method for 'coarse' positioning, and 'fine tuning' automatically occurs through the sequence specificity of the enzyme. In addition, with a proper choice of enzyme, cutting occurs with cohesive ends, which is suitable for ligation.

Figure 2.8 is a schematic representation of enzymatic molecular surgery on DNA. The DNA is stretched and anchored on the solid surface at the molecular ends. A DNA-cutting enzyme is immobilized on a microparticle, which is grasped by the optical pressure of a laser and moved to the target position. The particle is pressed against the stretched DNA so that enzymatic cutting occurs at the contact point.

An AFM stylus is an alternative probe, with which higher positioning accuracy may be expected. However, such mechanical probes have limited accessibility. DNA is stretched by the electrostatic method in a thin water layer sandwiched between a microscope slide and a cover slip. The cover slip must therefore be removed to allow access to the mechanical probe, and DNA strands are likely to be broken by hydrodynamic shear. In contrast, if a

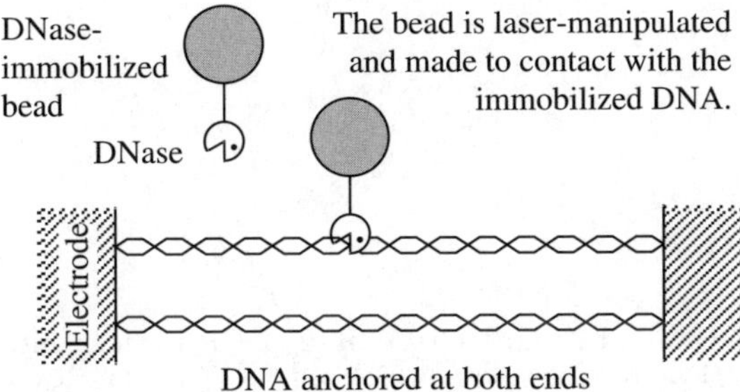

Fig. 2.8. Enzymatic molecular surgery

microparticle is used as the probe, it can be fed in from one edge of the cover slip and manipulated through the glass by a laser. There is then no need to remove the cover slip.

Immobilization of DNA. Proper contour fitting between the active site of the enzyme and the substrate (the target molecule) is a precondition for enzymatic activity. This imposes two requirements on DNA immobilization:

- The active site of the enzyme is often a pocket-like cleavage, represented by a mouth in Fig. 2.9. If the target DNA is fully adsorbed on the solid surface, the enzyme cannot accept the DNA deeply enough and will not be able to work. The DNA must therefore be held with a certain clearance above the solid surface, at least at the working point.
- Stretching may hamper enzymatic activity by deforming the DNA's helical structure, or by reducing its freedom to bend. Stretching should be somewhat loose in order to avoid too much tensile stress.

To meet these requirements, a microsystem for DNA immobilization has been developed, as illustrated in Fig. 2.10. It consists of a set of vacuum-evaporated aluminum electrodes, of which the two outermost energization electrodes are connected to a power supply, while the narrow strip electrodes in the gap have no electrical connections and are left at a floating potential. The function of

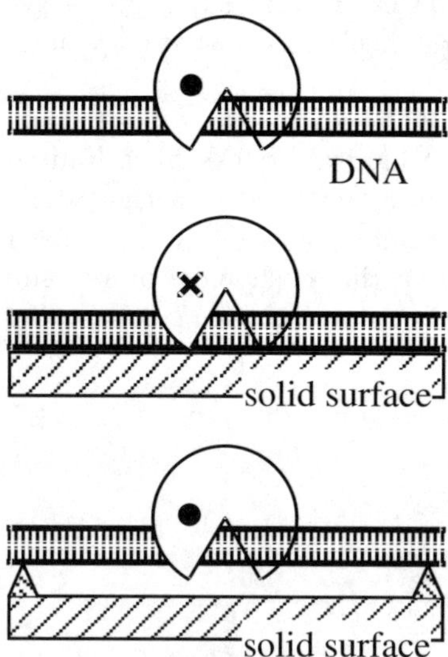

Fig. 2.9. Possible steric hindrance caused by immobilization (**a**) DNA must fit into the active site of DNase. (**b**) If DNA is fully adsorbed on a solid surface, DNase is unable to interact. (**c**) DNA must be held with some clearance so that the enzyme can bind. DNA should be held only at both ends.

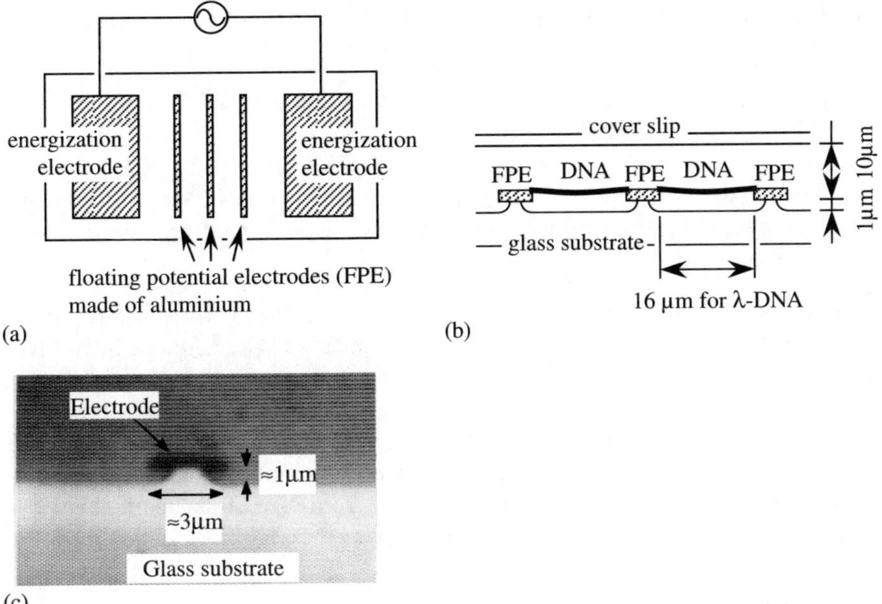

Fig. 2.10. Floating potential electrode system. (**a**) Top view, (**b**) enlarged cross-sectional view, (**c**) photo of the cross section

these floating potential electrodes is to deform the field pattern in such a way as to create field maxima at the electrode edges. When a voltage is applied to the energizing electrodes, DNA stretching occurs. One molecular end of the DNA is dielectrophoretically pulled onto the nearest electrode edge where the field maximum occurs, and anchored there. The spacing between the electrodes is made equal to the length of the DNA to be immobilized. The other end of the extended DNA therefore approaches the edge of the adjacent electrodes, where it is pulled in and anchored. We thus have DNA molecules anchored at both ends, bridging across two electrodes. By making the gap slightly smaller than the DNA length, we can obtain loose immobilization, which allows the DNA to bend.

As seen in the cross-section of Fig. 2.10b, the glass substrate just beneath the gap is etched down to form a trench with a depth of about 1 μm. Most DNA enzymes, including all restriction enzymes, require Mg ions for activity. Such multivalent positive ions may potentially cause adhesion of DNA onto a glass surface. The function of the trench is to ensure clearance and prevent the DNA from touching the solid surface.

The Probe. The probe we used for molecular surgery was a polystyrene microbead (Polybead, Polysciences Inc.) with a diameter of 1 μm. The enzyme is immobilized onto the bead using one of the following conventional biochemical procedures:

- The NH_2 on the enzyme is directly linked with COOH on the surface of the bead using EDC.[1]
- Avidin is immobilized on the bead using EDC, while the NH_2 or SH on the enzyme is labeled with biotin using biotin–BMCC,[2] and the two are mixed so that the enzyme and the bead are coupled by avidin–biotin binding.

The problem here is that the bead becomes very sticky with regard to the glass substrate when it has enzymes on the surface. We therefore introduced an anti-adsorption agent PEG-amine,[3] with a molar ratio of enzyme to spacer equal to 3:1. The resultant molecular composition on the surface is illustrated schematically for the second procedure in Fig. 2.11. PEG extending out from the surface of the bead plays the role of a bumper inhibiting adsorption, by preventing the bead (and the enzymes on it) from directly touching the glass surface. Making the density of the spacer sparse enough, DNA should still be accessible to the enzyme on the surface.

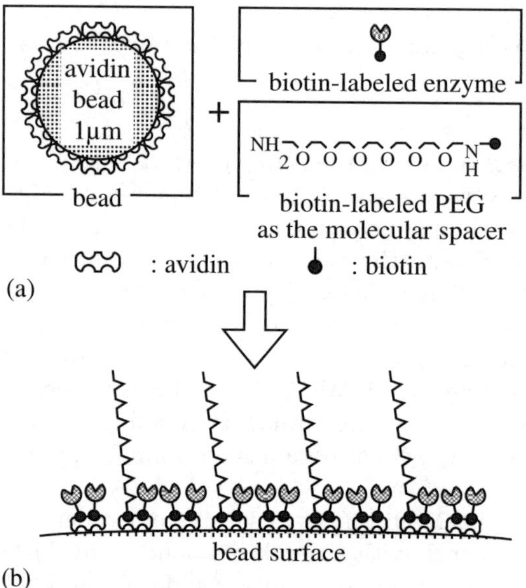

Fig. 2.11. Immobilization of the enzyme and the anti-stiction molecular spacer. (**a**) Reagents, (**b**) molecular structure on bead surface

[1] 1-ethyl-3(3-dimethyl-aminopropyl)-carbodiimide hydrochloride, Polysciences Inc.

[2] 1-biotinamido-4-[4′-(maleimidomethyl) cyclohexane-carboxamido] butane, Pierce Chemical Co.

[3] Poly(oxyethylene) dipropylamine, Wako Pure Chemical Industries Ltd., with molecular weight 10^4.

Experimental Results. The experimental demonstration of enzymatic molecular surgery was carried out using stretch-and-positioned λ-DNA. The enzymes used were two types of DNase, viz., DNase I and DNase II, both of which cut the DNA backbone regardless of the base sequence, and two restriction enzymes, viz., Hind III (6-base cutter) and Hha I (4-base cutter).

The DNA was stained with a fluorescent dye, Yo-Pro-1 (Molecular Probes Inc.), and observations were made under a fluorescent microscope. It was confirmed by a separate experiment that moderate staining with the fluorescent dye does not hamper enzymatic cutting. 10 μl of the DNA solution (60 pg DNA/ml) was fed onto the electrode system and covered with a 22 mm × 22 mm cover slip, resulting in an average solution thickness of 20 μm.

First of all, a control experiment was set up with a bead having no enzymes on the surface, hereafter referred to as the plain bead. Now both molecular ends of the DNA were anchored so that, if the bead was pressed against the DNA and moved far enough in a direction perpendicular to the strand until the tension in the DNA exceeded its strength, the strand would break mechanically. The DNA deformation at this point was used as a reference in the later experiment to determine whether cutting was enzymatic or mechanical.

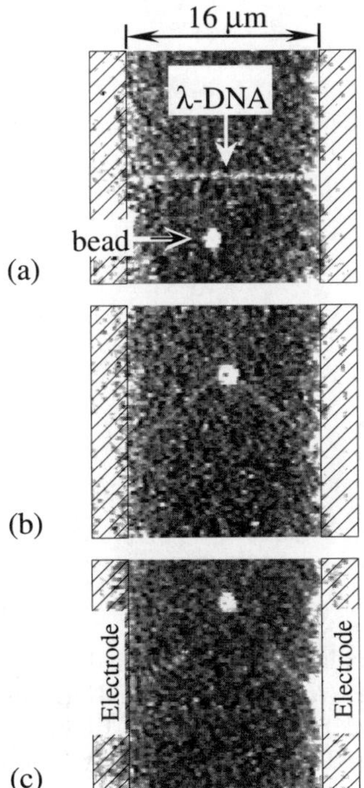

Fig. 2.12. Mechanical breakage of DNA. (**a**) Immobilized DNA and a plain bead. (**b**) λ-DNA originally 16 μm in length is mechanically stretched to 21 μm. (**c**) The DNA breaks and the fragments shrink back to the electrodes

Figure 2.12 shows the result. The bead shown by an arrow was laser-manipulated and pressed upward against the DNA in the figure. (As a matter of fact, the bead in this photo is a fused agglomerate of two beads. A spherical bead slips when it is pressed against DNA, and cannot be pushed hard enough.) No cutting was observed by a mere touch of the bead. When the bead was moved further upward, the DNA was bent and mechanically elongated, but still no cutting occurred (Fig. 2.12b). When moved still further (Fig. 2.12c), mechanical breakage occurred. It was observed that breakage always occurred at the contact point with the bead, and never at the anchored point or the middle, although the reason for this is not clear. The average elongated length of the DNA when breakage occurs is as long as 25 μm, which is 150% of the full length of the DNA.

Figure 2.13 shows molecular surgery with the DNase-I-labeled bead. Figure 2.13a is a frame before cutting. It shows the bead and an immobilized DNA strand bridging the electrodes. Figure 2.13b shows the moment when the bead has just touched the DNA. Figure 2.13c is the next video frame (shot 30 ms later), in which the DNA is already cut and the cut molecule is shrinking back to the electrode. The same experiment was repeated and in all experiments DNA cutting was observed just as the bead touched

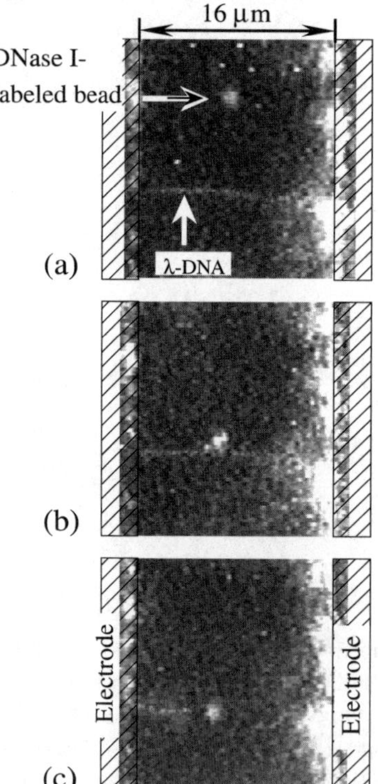

Fig. 2.13. Molecular surgery with a DNase-I-labeled particle. (**a**) Immobilized DNA and a DNase-I-labeled bead. (**b**) DNase-I-labeled bead just touches DNA. (**c**) The DNA is instantaneously cut and the left half of the molecule shrinks back towards the electrode

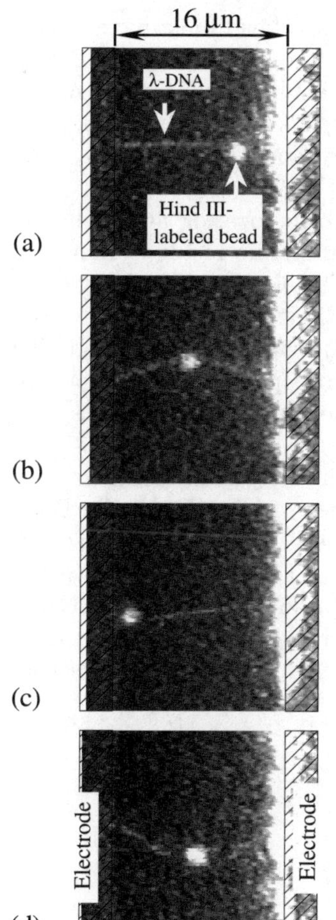

Fig. 2.14. Molecular surgery with a restriction-enzyme-labeled particle. (**a**) Immobilized DNA and a restriction-enzyme-labeled bead. (**b**) The bead is moved upwards and the DNA is bent but not cut. (**c**) The bead is moved leftward along the DNA and reaches the left end without cutting. (**d**) The bead is moved back (towards the right) and the DNA is finally cut

the DNA. Experiments with the DNase II bead showed the same phenomena.

The result of the experiment using the 6-base cutter, Hind III, is shown in Fig. 2.14. When the Hind-III-immobilized bead is pressed against stretch-and-positioned DNA, two cases are observed. One is rarer than the other, in which the DNA is cut as soon as the bead touches it. This was observed in only a few percent of the trials. The other is the dominant case, in which the DNA is not instantaneously cut. Figure 2.14 represents the second case. In Fig. 2.14a, the Hind-III-labeled bead is already in contact with DNA, but the DNA remains intact. First the bead was moved perpendicularly to the strand (but not so far as to cause mechanical breakage), in order to press it more firmly against the DNA (Fig. 2.14b). The DNA remained in place. Then the bead was moved leftward (relative to the photo) to the end of the DNA strand, whilst maintaining contact (Fig. 2.14c). Because Hind III has 7 restriction sites on λ-DNA, the enzyme-labeled bead must have passed several restriction sites, but cutting did not take place. To continue scanning, the bead was then moved back to the right. At some point, the DNA was suddenly cut (Fig. 2.14d). As typically shown by this case, cutting does not take place by a mere touch, but requires scanning along the DNA. The same thing is observed with the Hha-I-labeled bead.

Repeating the same experiment, we examined the location at which cutting by the restriction-enzyme-labeled bead occurs. The result is shown in Fig. 2.15 in comparison to the restriction map. The electrostatically stretch-and-positioned DNA is a mixture of two orientations, so the restriction positions for the two orientations are also shown in the figure. Although the accuracy of this experiment using an optical microscope is limited by the optical resolution, it clearly shows peaks in cutting positions, which correspond well to the restriction map.

The reader should note the following experimental finding. It was expected to be difficult to maintain contact between the bead and the DNA during scanning, but this was not actually the case. The bead was observed to be somewhat sticky with respect to the DNA. This may be due to the nature of the DNA enzymes, because such non-specific binding with DNA is advantageous in seeking the restriction site. If there were no such interaction, the enzyme would have to rely on 3D diffusion until it finally hit the restric-

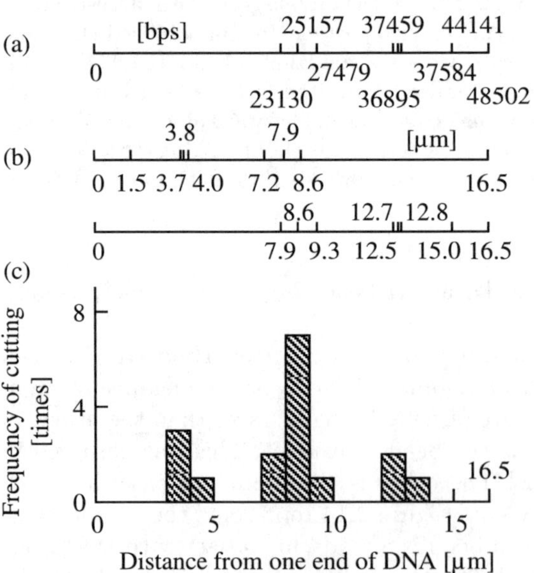

Fig. 2.15. Location of observed cutting with Hind-III-labeled particle. (**a**) Restriction map of λ-DNA (in base pairs). (**b**) Restriction map in actual length from one end of the DNA for the two possible orientations (in μm). (**c**) Cutting frequency as a function of distance from one end of the DNA (in μm)

tion site by chance. On the other hand, with this interaction, the enzyme can first bind to any location on the DNA and slide in a one-dimensional manner along the strand, whereby it can more easily locate the site [7]. The techniques developed here may find application in the direct measurement of such interactions.

2.2.4 Use of Local Temperature Rise

Another method for inducing space-resolved chemical reactions on a stretched DNA is local temperature rise, in which the temperature dependence of chemical reactions can be exploited. For instance, if the temperature of a particular portion of the DNA is thermally-cycled by a microheating device, it should be possible to copy the local sequence (local PCR). Heating can be carried out by microheaters installed on the substrate, or a movable heater that can be manipulated and brought into contact with the strand.

The idea of local temperature rise is demonstrated using bacterial flagella as a specimen. A flagellum is a bacterial tail, consisting of non-covalently bound protein sub-units called flagellins that dissociate above 60°C. It can thus be used as a temperature rise indicator. A 2.8 μm diameter latex particle containing fine iron powders was used as the microheater. The particle has a moderate absorbance and can thus be laser-trapped, whilst simultaneously generating a temperature rise.

Figure 2.16a1 shows a flagellum and a bead under a fluorescent microscope. When the bead is laser-manipulated and contacted with the flagellum, it is cut there (Fig. 2.16a2). With reduced laser power, the flagellum is not cut but just bent (Fig. 2.16b1, 2).

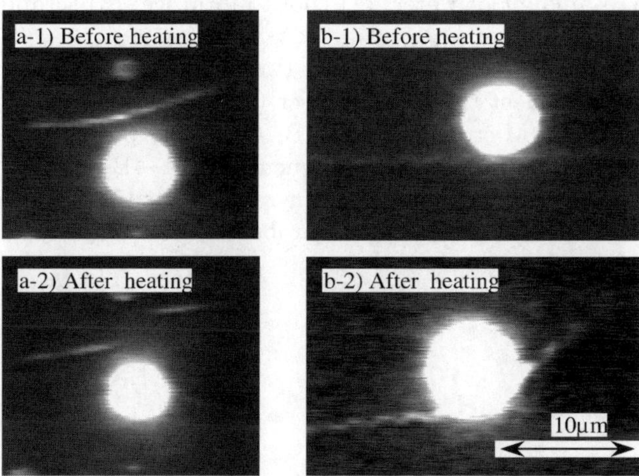

Fig. 2.16. Local conformation change and dissociation of a flagellum by pinpoint laser heating. (**a**) Cutting a flagellum. (**b**) Transforming a flagellum

2.3 A Microfabricated Probe for Molecular Surgery

In the last section, we discussed molecular surgery using enzyme-immobilized particles with diameter 1 μm. We may expect higher positional accuracy by using a smaller particle. However, there are difficulties:

- as the particle becomes smaller, laser trapping becomes more difficult, because the optical trapping force is smaller and thermal randomization is greater,
- the smallest size of the focal spot is about the wavelength of the laser beam, and when smaller particles are used, many particles are pulled into the spot and simultaneously trapped there.

In order to achieve high spatial resolution, we used microtechnology to fabricate a probe for laser manipulation. This is shown schematically in Fig. 2.17. The probe consists of a base plate with a sharp tip on which enzymes are to be immobilized, and more than three spherical microparticles attached on the plate. By grasping and controlling the position of each bead with a laser, one can translate or rotate the probe. The device achieves a high resolution as determined by the sharpness of the tip, and stable laser trapping is provided by relatively large spherical particles.

The fabrication of the device starts with a cantilever-like structure made from an SiO_2 layer deposited on a silicon wafer (Fig. 2.18). Then the structure is labeled with amino groups and the amino-labeled latex bead is bonded onto it using molecular linkers such as glutaraldehyde. Molecular patterning can be used to define the location of the microparticles, but in the present experiment, they are randomly attached. The device thus formed on the Si substrate is removed from the substrate and isolated by a freeze-fracture-type process: (1) apply a small amount of water, (2) cool to form ice, (3) scrape off, and (4) melt.

The optical system used in the experiment had two computer controlled galvano-scanning mirrors to deflect the focused laser (1 W YAG) in the xy plane of the microscope view. Microparticles A, B, C on the probe were scanned by the laser on a time-sharing basis in a sequence like A→B→C →A, with 5 ms each. Moving the laser but maintaining the relative positions of the three particles, the position and orientation of the probe were controlled. The

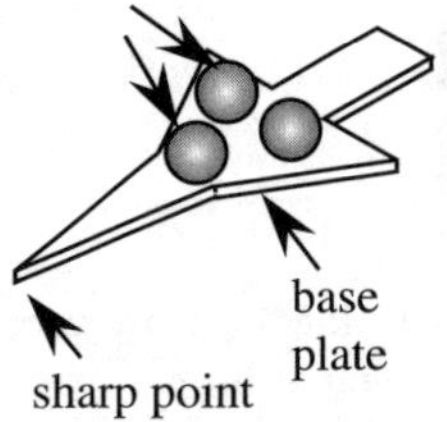

Fig. 2.17. Microprobe for laser manipulation

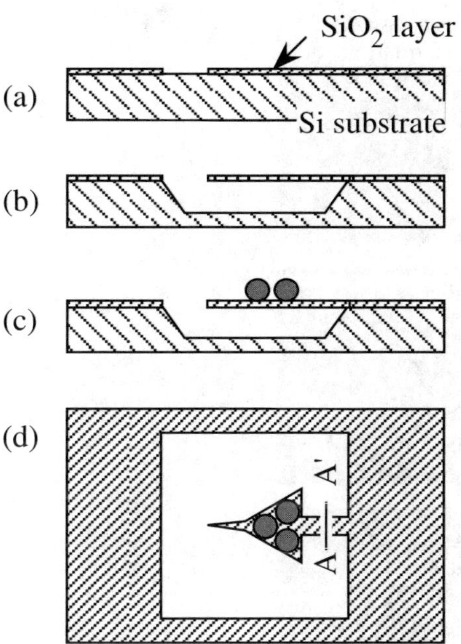

Fig. 2.18. Fabrication of the microprobe. (**a**) Patterning (cross-sectional), (**b**) formation of cantilever structure, (**c**) attachment of latex particles of diameter 3 μm, (**d**) breaking at AA′ and recovery (plan view)

calculated amplitude of vibration at the tip resulting from the time-sharing scheme is several tens of nm for the given probe geometry and experimental conditions.

Figure 2.19 shows snapshots of the translation and rotation of the probe as they were actually realized in a liquid with specific density 1.2. The probe was made to follow a square trajectory as depicted in Fig. 2.19f, always keeping the tip in front. A translation speed of about 10 μm/s could be achieved. This was limited by the trapping force of the laser.

2.4 Conclusion

Microfabrication is a powerful tool for bio-nanotechnology. Molecular surgery, that is, pinpoint modification on the single-molecule level through physical manipulation, has now become a reality. Several examples are reported here, taking DNA as the target. Electrostatic field effects are used to stretch and immobilize DNA at a predetermined position on a solid surface, and a laser, a mechanical stylus, or an enzyme-immobilized probe is used to cut at the target position on the stretched strand.

The potential applications of molecular surgery are not limited to cutting. With the use of an enzyme-immobilized probe, various molecular modifica-

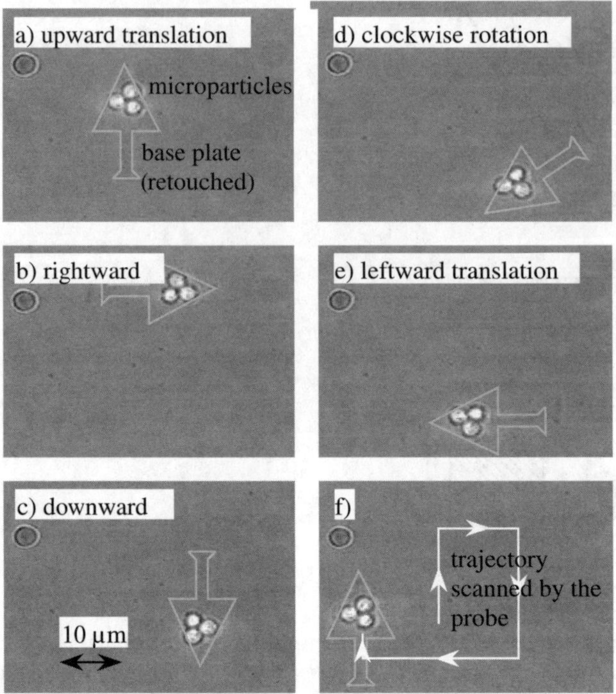

Fig. 2.19. Translation and rotation of the probe

tions should be possible. Basic investigations of the molecular interaction between DNA and DNA enzymes through spatially and temporally resolved observation may be another area of application [11].

The number of molecules handled in molecular surgery is far smaller than in conventional biochemistry. This is an advantage in that only a small amount of sample is required, but poses some difficulties in chemical assays and detections. If biochemistry based on molecular surgery is to be established, novel chemical processes involving very small numbers of molecules must be developed alongside. This is yet another challenge for microtechnology.

Acknowledgements. The author gratefully acknowledges collaboration and discussions with Mr. Osamu Kurosawa of Advance Co., Dr. Hiroyuki Kabata of Kyoto University, Dr. Takatoki Yamamoto of Tokyo University, Prof. Nobuo Shimamoto of the National Institute of Genetics, Dr. Keiichiro Okabe of Advance Co., and Dr. Seiichi Suzuki of Seikei University. This work was in part supported by the Bio-oriented Technology Research Advancement Institution (BRAIN Seiken Kiko, Shinjigyo Soshutsu Kenkyu Kihatsu Jigyo), NEDO (Sangyo Kagaku Gijutsu 97S07-005-2), the Ministry of Education (Grant-in-Aid for Scientific research: 11450103, 10175216), Advance Co., and the Toyota Physical and Chemical Institute.

References

1. H.A. Pohl: *Dielectrophoresis* (Cambridge University Press, Cambridge 1978)
2. T.B. Jones: *Electromechanics of Particles* (Cambridge University Press, Cambridge 1995)
3. R. Pethig: *Application of AC Electric Fields to the Manipulation and Characterisation of Cells*, in *Automation in Biotechnology*, ed. by I. Karube (Elsevier Science Publishers 1991) pp. 159–185
4. M. Washizu, O. Kurosawa: *Electrostatic Manipulation of DNA in Microfabricated Structures*, IEEE Trans. IA **26**, No. 6 (1990) pp. 1165–1172
5. M. Washizu, O. Kurosawa, I. Arai, S. Suzuki, N. Shimamoto: *Applications of Electrostatic Stretch-and-Positioning of DNA*, IEEE Transaction IA. **31**, No. 3 (1995) pp. 447–456
6. S. Suzuki, T. Yamanashi, S. Tazawa, O. Kurosawa, M. Washizu: *Quantitative Analysis on Electrostatic Orientation of DNA in Stationary AC Electric Field Using Fluorescence Anisotropy*, IEEE Transaction IA **34**, No. 1 (1998) p. 75–83
7. H. Kabata, O. Kurosawa, I. Arai, M. Washizu, S.A. Margason, R.E. Glass, N. Shimamoto: *Visualization of Single Molecules of RNA Polymerase Sliding along DNA*, Science **262**, 1561–1563 (1993)
8. O. Kurosawa, K. Okabe, M. Washizu: *DNA Analysis Based on Physical Manipulation*, Proceedings of the 13th Annual International Conference on Micro Electro Mechanical Systems (MEMS 2000) pp. 311–316
9. A. Bensimon, A. Chiffaudel, V. Croquette, F. Heslot, D. Bensimon: *Alignment and Sensitive Detection of DNA by a Moving Interface*, Science **265**, 2096–2098 (1994)
10. T. Yamamoto, O. Kurosawa, H. Kabata, N. Shimamoto, M. Washizu: *Molecular Surgery of DNA Based on Electrostatic Micromanipulation*, Conf. Rec. 1998 IEEE/IAS Annual Meeting 44-03 (1998) pp. 1–8
11. H. Kabata, W. Okada, M. Washizu: *Single-Molecule Dynamics of the Eco RI Enzyme Using Stretched DNA: Its Application to in Situ Sliding Assay and Optical DNA Mapping*, Jpn. J. Appl. Phys. **39**, 7164–7171 (2000)

3 Manipulation of Single DNA Molecules

Akira Mizuno

In the post-genome stage, DNA information concerning individual people can be used for protection against or treatment of diseases. To meet this requirement, more rapid DNA analysis methods will be needed. Analytical methods based on single DNA molecules may replace the conventional DNA analysis procedure. This method also provides a more precise tool for studying interactions between DNA molecules and protein molecules. Single DNA can be prepared by the following operations: manipulation of genomic DNA, fixation in stretched form, mapping, cutting, recovery of the fragment, and PCR amplification. These operations can be conducted in a very small integrated analytical system made on a silicon substrate. The globular transformation can avoid breakdown of long genome DNA and permits manipulation of large DNA. Because the globular transition is reversible, the DNA can be pulled out sequentially from the globule of DNA and fixed on a substrate in an arbitrary pattern. This DNA molecule can be characterized by attaching a restriction enzyme or by probes. Eliminating the activity of the restriction enzyme by removing Mg ions, fluorescence-labeled EcoRI is attached and an optical restriction map can then be made. Releasing Mg ions locally can then cut the DNA. The fragments can be recovered on a glass capillary using electrophoresis. The PCR method starting from an unknown single fragment of DNA molecule has been developed using a microreactor system based on a water-in-oil emulsion. Using this system, single DNA can be amplified. This system is suitable for manipulating and reacting single DNA molecules in an integrated analytical apparatus on a silicon substrate.

Single-molecule techniques provide a new analytical method differing completely from conventional analytical methods in biology and biotechnology. The analysis based on fluorescence observation supports real-time observation in a liquid environment. Brownian motion of individual DNA molecules has been observed using a fluorescence microscope [1]. The motion of an actin filament over heavy meromyosin has been analyzed by observing a single fluorophore bound on the actin [2]. A T2 phage DNA molecule driven by a pulsed field in agarose gel has been visualized, and this visualization has contributed to investigations concerning the mechanism of DNA separation by gel electrophoresis [3]. RNA polymerase–DNA complexes have been observed and the transcription force with which a single RNA polymerase pulls the DNA molecule [4] or slides along the DNA molecule array [5] have been studied.

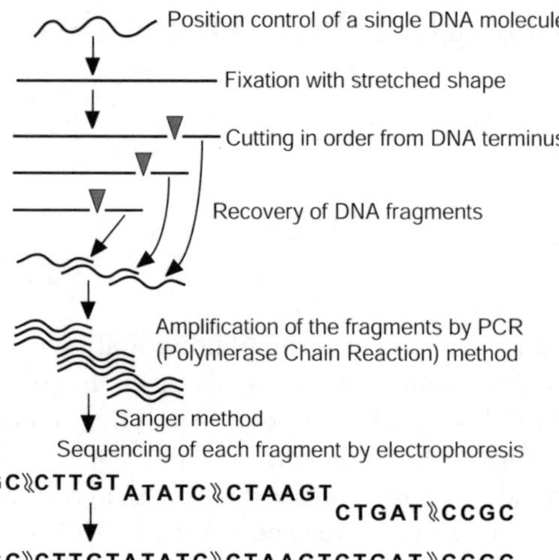

Fig. 3.1. Illustration of genome sequence determination based on successive cutting and recovery of DNA fragments. It is not necessary to identify the overlap region

A manipulation technique for single DNA molecules opens the way to new analysis features, such as mapping and sequencing. Conventional DNA sequencing methods analyze only about 1 000 base pairs from one sample at any one time. A long DNA, such as genetic DNA, must therefore be cut into many short analyzable fragments by one of several methods. However, these fragments lose information concerning their order and a reorganization process is required to determine the whole sequence after each fragment has been analyzed. This process requires time and effort. If fragments of single DNA can be prepared without losing order information, the speed with which genetic DNA can be sequenced will be enhanced. Figure 3.1 shows the idea underlying a novel method for preparing fragments without loss of order information. The method consists of the following operations:

- fixation of a single DNA molecule with stretched shape,
- mapping and cutting,
- successive recovery of the fragments,
- amplification of the recovered DNA fragment by PCR (polymerase chain reaction).

In single-molecule PCR, fine droplets can be used as a container for molecules and chemicals. The speed of the chemical reaction can be increased through the reduced distance between molecules that are to react together. This single-molecule system requires only a very small amount of sample and chemicals and can be integrated into a microfluidic system or micro-total analytical system (μ-TAS) constructed using semiconductor technology. In this chapter, we describe the above operations in order to present the current state of real-time manipulation of single DNA molecules.

3.1 Manipulation of Giant DNA Molecules

Preparation and manipulation of giant DNA molecules has been recognized as an important technique for biochemistry and molecular biology [3–9]. They also constitute an important step in genome analysis based on single-molecule DNA. For the manipulation of single DNA molecules, both laser trapping and electrostatic force are major tools, as shown in Fig. 3.2. Laser trapping, or laser tweezers, developed by Ashkin [10–12], are a useful tool for biological research [13–15], because the laser beam can be focused to about the size of its wavelength and suspended particles can be trapped without physical contact. A DNA–bead complex can be optically trapped and manipulated using a YAG laser [7]. Chu et al. [8] attached one terminus of a lambda phage DNA molecule and trapped it optically. Electrostatic force is also useful for manipulation. Recent developments in photolithographic methods make it possible to generate a very high and accurate electric field in micro-fabricated apparatus. Electrostatic effects such as electrophoresis and dielectrophoresis have been applied to orientate or stretch DNA molecules [16,17]. A combination of laser trapping and electrostatic force enables more flexible micromanipulation of cells and DNA molecules [9].

Even though electric forces can be applied to manipulate small DNA molecules, large chromosomal DNA molecules (more than several hundred kilo base pairs) in solution are difficult to manipulate, because large DNA molecules are easily broken down by shear stress in the flow. To suppress fragmentation of DNA due to shear stress, the globular transition can be utilized. Because the globular transition induces high condensation on DNA, the transition also permits manipulation of giant DNA, such as yeast chromosomal DNA, by direct laser trapping.

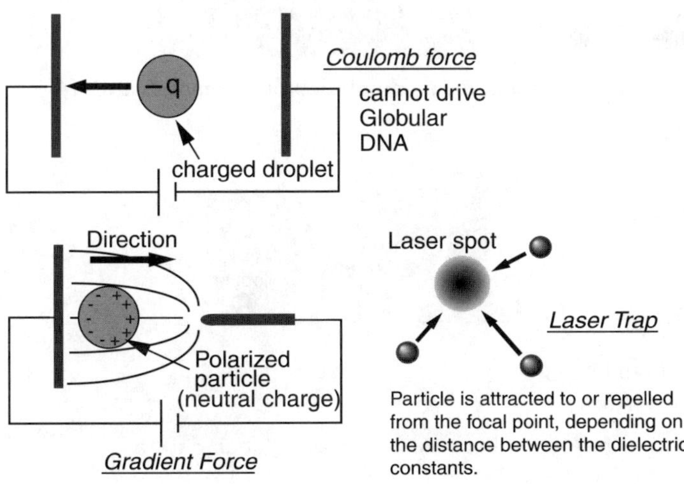

Fig. 3.2. Non-contact manipulation by electrostatic force/laser trapping

3.1.1 Characteristics of Globular DNA

Many studies have been made on DNA condensation using condensing agents [18–20]. For example, PEG (polyethylene glycol) in combination with low molecular weight inorganic salt induces DNA to transform from the coiled to the globular structure, as shown in Fig. 3.3. The transformation is referred to as polymer-and-salt-induced-phi condensation. The globular DNA string is arranged as a hexagonal, close-packed structure [21,22]. Phi condensation can be induced by several kinds of polymers and salts. It has been reported that the conformation of DNA can be changed from the coiled to the globular state with increasing concentration of PEG or salt, as shown in Fig. 3.4. This condensation is reversible, and removal of PEG and/or salt in solution leads to transformation from globular to coiled DNA [23]. This is an important characteristic of the transformation to be used for DNA manipulation.

The globular transformation is rapid. Figure 3.5 shows the transformation of a coiled T4 phage DNA to a globule in less than 2 s. The DNA shrinks with the growth of nucleation. Polyethylene glycol and chlorine ions induce this transformation. PEG and $MgCl_2$ induce the toroidal structure as shown in the TEM photographs (Fig. 3.5b).

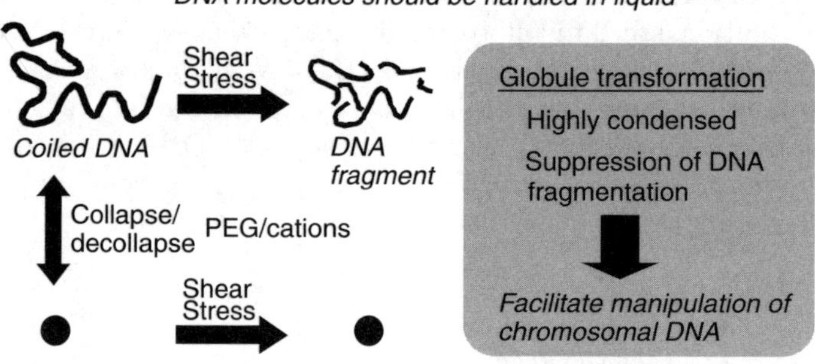

Fig. 3.3. Coil–globule transformation for manipulation of chromosomal DNA

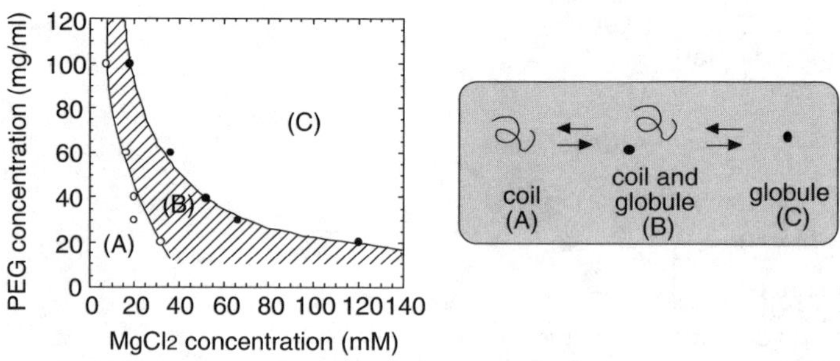

Fig. 3.4. Reversible transformation between coiled and globular state

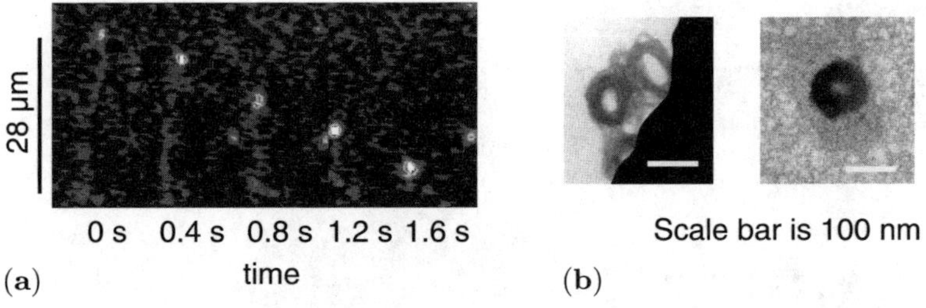

Fig. 3.5. Transformation of T4 DNA to globular state. Condensing reagents: PEG/$MgCl_2$, spermidine. (**a**) Growth of nucleation from one end. (**b**) TEM images of globular DNA

3.1.2 Suppression of Fragmentation by Globular Transition

The toughness of globular DNA against shear stress was examined with yeast chromosomal DNA prepared by the following procedure [24]. Yeast (Saccharomyces cerevisiae) cells (S288C strain) were cultured in YPD solution [5 g/l yeast extract (Nacalai Tesque), 10 g/l polypeptone (Wako), 2% glucose (Nacalai Tesque)] at 30°C for 3 days. The cultured yeast cells were centrifuged at 3 000 rpm for 5 minutes and the precipitates were washed once with 50 mM EDTA solution and twice with CES solution [0.5 M sorbitol (Nacalai Tesque), 0.1 M sodium citrate (Nacalai Tesque), 50 mM EDTA]. The precipitates were mixed with a melted low-melting-point agarose (Gibco BRL) and the mixtures were solidified in 50 µl of agarose plug. The plug was treated with 0.14% zymolyase (Seikagaku) at 37°C for 3 hours. The next stage in the treatment was deproteinization with 0.5 mg/ml Proteinase K (Promega) at 50°C overnight. For the transition to a globular structure, the gel plugs containing the chromosomal DNA were soaked overnight in a solution of 60 mg/ml PEG and 0.6 M NaCl. The extracted DNA was stained with a fluorescence dye YOYO-1 iodide (Molecular Probes) for visualization.

The toughness of the globular DNA was examined by observing the DNA shape before and after subjecting to shear stress. A gel plug containing the globular DNA on a cover slip was heated to about 70°C (the melting point of agarose gel) and pressed as shown in Fig. 3.6. An inverted fluorescence microscope was used to observe the melted gel plug. Figures 3.7a and b show microscope images of random coil and globular chromosomal DNA molecules, respectively. As shown in Fig. 3.7b, DNA molecules were condensed by transition to globular structure. These molecules were treated with a shearing stress generated by melting the agarose gel. DNA images after the treatment are shown in Figs. 3.7c and d. In the melted gel plug, the random coil DNA molecules were significantly stretched by the shearing stress (see Fig. 3.7c). On the other hand, globular DNA molecules kept the same structure (see Fig. 3.7d). This result suggests that globular DNA is toughened against shearing stress, whist fragmentation is suppressed.

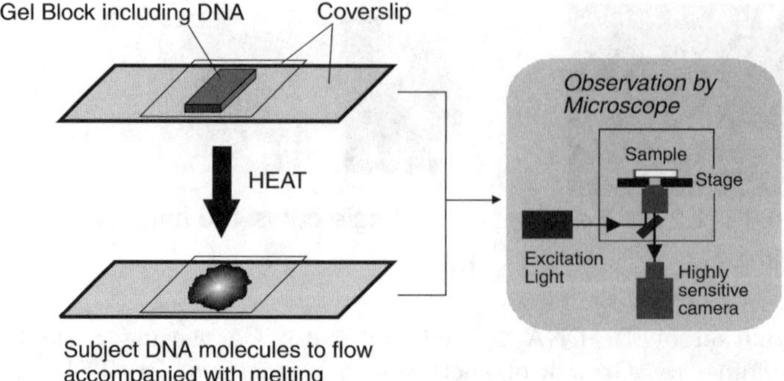

Fig. 3.6. Observation of the effect of shear stress on DNA in agarose gel

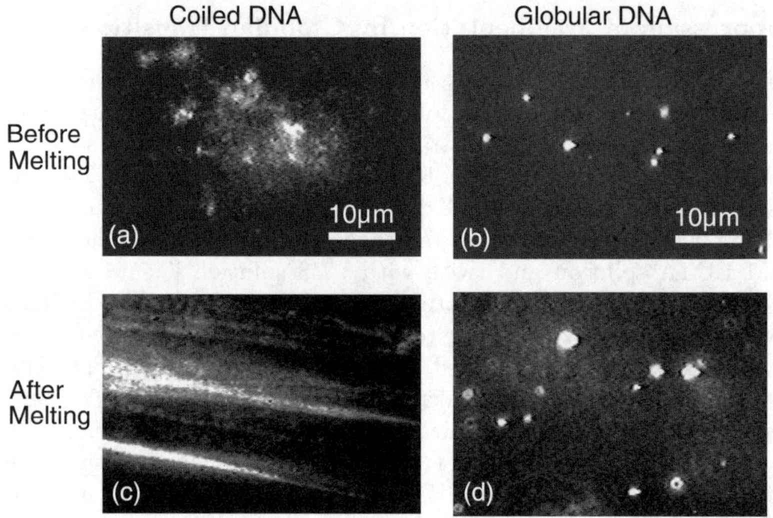

Fig. 3.7. Photographs of coiled and globular DNA before and after exposure to shear stress by melting a gel

We further investigated the effect of shearing stress on globular chromosomal DNA using pulsed-field gel electrophoresis. A gel plug containing globular DNA or random coil DNA (non-globular DNA) was prepared. Both samples were melted in a microtube and treated with a vortex mixer (Vortex Genie 2, Scientific Industries) for 10 s with different strengths: weak stress corresponded to 3 on the vortex mixer dial and strong stress to 6. These chromosomal DNAs were analyzed by a pulsed-field gel electrophoresis system. Figure 3.8 shows pulsed field gel electrophoresis analyses for a random coil (lanes a–c) and globular yeast chromosomal DNA molecules (lanes d–f) treated with the Vortex Genie 2 (vortex mixer). Lanes a and f denote chromosomal DNA with no treatment. Lanes b and d denote DNA molecules treated

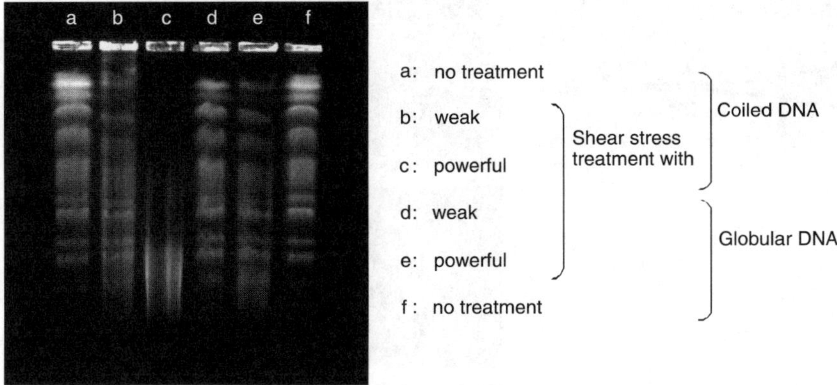

Fig. 3.8. Pulsed-field electrophoresis of yeast chromosomal DNA with and without shear stress applied by vortex mixer

for 10 s with the Vortex Genie 2 on dial setting 3. Lanes c and e denote 10 s treatment with the Vortex Genie 2 on dial setting 6, which caused much higher shearing stress than dial setting 3. Shorter DNA fragments observed in lanes b and c demonstrated that random coil DNAs are easily fragmented by shearing stress. Although random coil DNAs were fragmented (lanes b and c), transition to globular structure suppressed fragmentation due to shearing stress (lanes d and e).

These results demonstrate that globular transition suppresses fragmentation of giant DNA. This means the globular DNA can be manipulated without fragmentation caused by shear stress. A glass capillary can be used to handle globular DNA. Figure 3.9 is a sequence of photographs showing the recovery by a capillary of a single globular DNA molecule from a liquid. The recovered single DNA molecule clearly maintained its structure in the fluid during the recovery process.

Figure 3.10 shows the method for preparing globular DNA. As shown in Fig. 3.10a, yeast cells were treated in agarose gel to avoid fragmentation and chromosomal DNA was recovered in the gel. The gel plug was placed in the electrophoresis apparatus shown in Fig. 3.10b. The large DNA molecules were driven towards the well containing agents for globular transformation. A membrane was used to block the DNA molecules for recovery.

Chromosomal DNA from rats can also be prepared as a globule. For preparation, the cell cycle should be synchronized and the M-stage cell is processed to extract DNA in the gel plug. Using the apparatus in Fig. 3.10b, DNA is recovered as a globule.

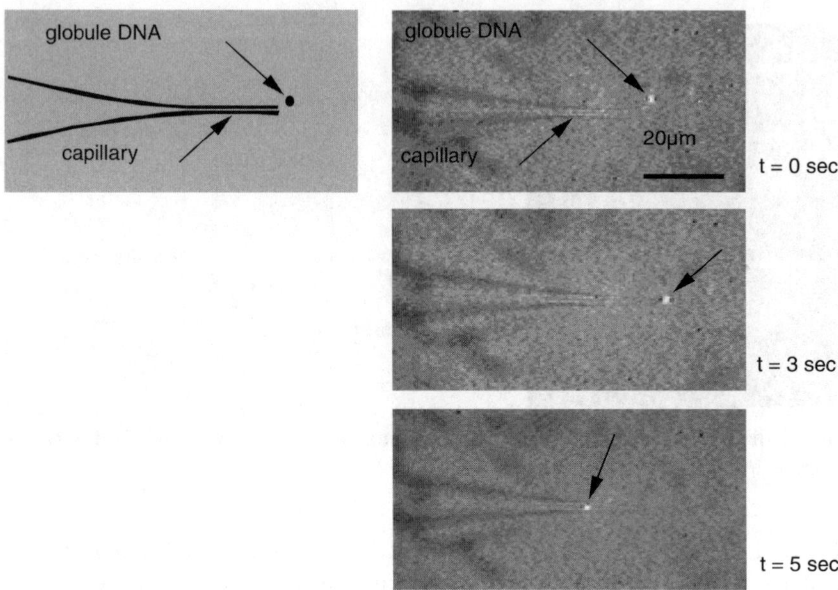

Fig. 3.9. Recovery of globular DNA using a glass capillary

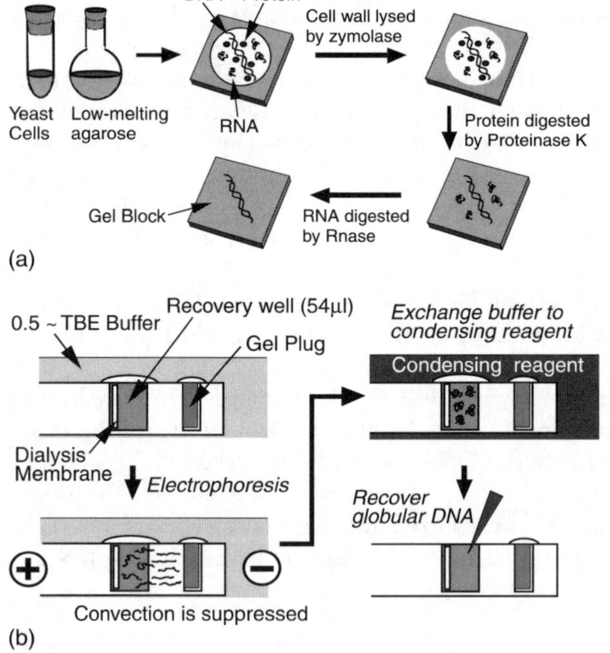

Fig. 3.10. Preparation of globular DNA. (**a**) Extraction of chromosomal DNA. (**b**) Cross-section of recovery apparatus for globular chromosomal DNA

3.1.3 Laser Trapping of Single DNA

Another single DNA manipulation technique is indirect laser trapping using a DNA–bead complex [25]. A μm-sized bead can be attached to one end of a DNA molecule and manipulated using laser trapping. The DNA–bead complex is depicted in Fig. 3.11. DNA molecules are bound to a 1 μm latex bead through the specific interaction between biotin and avidin. Avidin is a tetrametric protein with the capacity to bind four biotin (vitamin H) molecules with a dissociation constant of 10–15 M, and this binding is well known as one of the strongest non-covalent interactions. The terminus of the T4 DNA molecule is labeled with a dUTP analogue bound to the biotin by a T4 DNA polymerase reaction (3′ to 5′ exonuclease and replication activity) [40] and then bound to avidin-coated 1 μm latex beads.

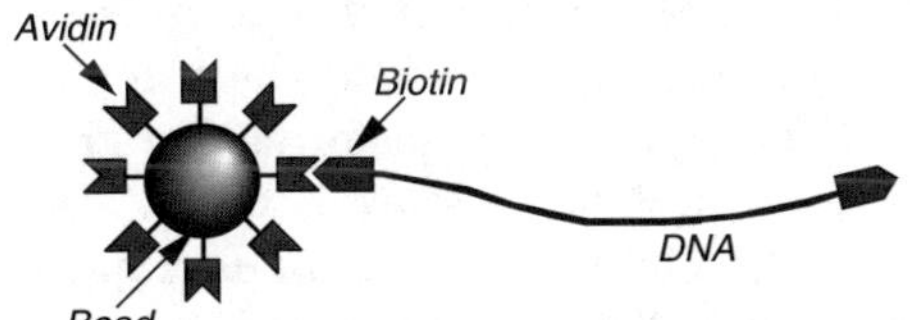

Fig. 3.11. Bead–DNA complex

The biotin-labelling procedure for the DNA ends is as follows:

1. T4 DNA (25.6 μl in 320 μl reaction buffer).

   Reaction buffer: 50 mM Tris-Cl (pH 8.5), 7 mM $MgCl_2$,
   15 mM $(NH_4)_2SO_4$, 10 mM 2-mercaptoethanol,
   0.1 mM EDTA, 64 U T4 DNA polymerase.

2. Incubate at 37°C for 5 min.
   T4 DNA polymerase digests the DNA strand from the 3′ to the 5′ terminus. This reaction creates a sticky endpoint to be filled by a polymerization reaction.
3. Add 16 μl of 1 mM d(ACG)TP, 16 μl of 0.5 mM biotin-21-dUTP.
4. Incubate at 37°C for 60 min.
   T4 DNA polymerase synthesizes DNA at the sticky endpoint.
5. Stop reaction by adding 16 μl of 0.5 M EDTA.
6. Incubate at 65°C for 10 min to deactivate T4 DNA polymerase.
7. Dialyze against TE.

Figure 3.12 shows a trapped DNA–bead complex inside a glass capillary. The single DNA molecule was fixed with 1 μm bead by laser trapping and stretched by solution flow and/or electrophoresis.

The transformation to the globular state has advantages not only for improved strength against shear stress but also for manipulation by laser trapping. A sharply focused laser can trap small particles at the laser focal point due to the difference in refractive index between the particles and

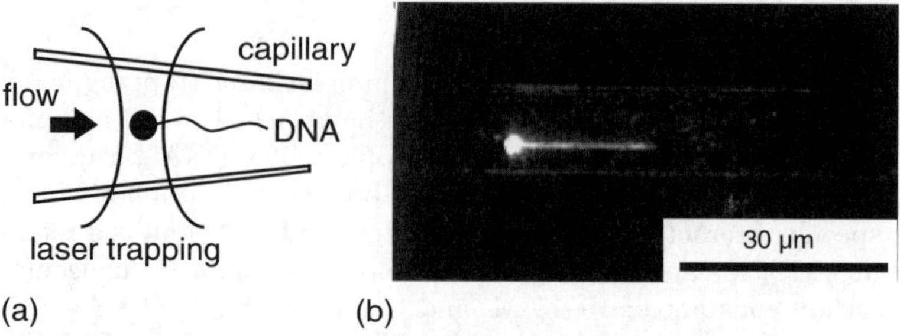

Fig. 3.12. Laser trapping of a bead–DNA complex, stretched by flow in a capillary. (**a**) Schematic diagram. (**b**) Photograph

the surrounding medium. However, it has proved difficult to trap biological macromolecules such as DNA molecules in aqueous solution with low viscosity, because these molecules differ little in refractive index and the trapping force cannot overcome Brownian motion. The globular transformation gives the DNA a higher refractive index, because the globule state is highly condensed and has increased density. The globular DNA molecule can be trapped optically using an Nd:YAG laser (wavelength 1 064 nm). It should be noted that coiled DNA might be optically trapped if Brownian motion were suppressed by increasing the viscosity using PEG.

Figure 3.13 depicts the experimental apparatus for DNA manipulation by laser trapping. Fluorescence-stained DNA molecules were observed using an inverted microscope equipped with a 100×, 1.3 NA oil-immersion objective lens and a high sensitivity SIT camera. Images were recorded on video tape. The 1 064 nm Nd:YAG laser beam, expanded to a diameter of 7.5 mm, was introduced into the microscope through a dichroic mirror. The beam was focused on the objective plane of the fluorescence image by the objective lens in order to trap DNA molecules optically. Approximately 180 mW (measured

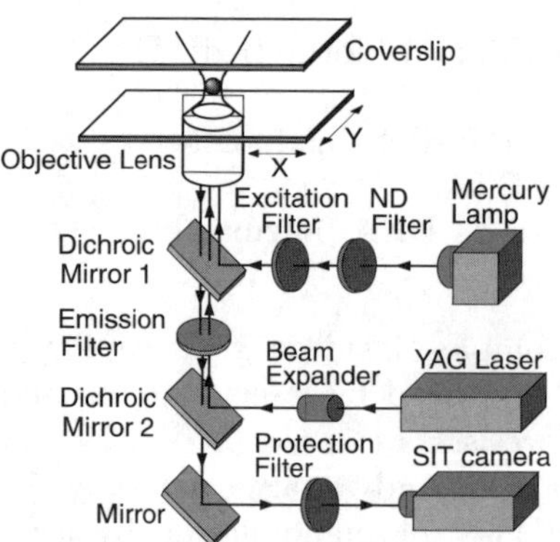

Fig. 3.13. Optical system for laser trapping

at the back of the objective lens) of 1 064 nm light was used to trap globular DNA.

Globular DNA induced by PEG–$MgCl_2$ was optically trapped using 180 mW of 1 064 nm light from an Nd:YAG laser. The infrared laser causes less damage to biological molecules such as DNA than visible light from krypton ion or argon ion lasers [26]. Before trapping the globular DNA, the transition was confirmed by fluorescence microscope. Under the conditions [PEG] = 60 mg/ml and $[MgCl_2]$ = 50 mM, all DNA molecules were in the globular state and no coiled DNA was observed. Figure 3.14 shows laser trapping of a globular DNA molecule. Free DNA molecules were translated by moving the microscope stage. The globular DNA molecule indicated by white arrows was trapped at the laser focal point while a free DNA molecule moved leftward (Figs. 3.14a–c) or upward (Figs. 3.14c–e). At this laser power, it was possible to trap the globular DNA. However, coiled DNA molecules could not be trapped. When a coiled DNA molecule arrived at the laser focal point by moving the stage, it was pushed downward by the optical pressure of the laser beam and escaped from the trap. Figure 3.15 shows the trapping force on a globular T4 phage DNA. The trapping force becomes stronger with increased laser power and increases with fragment size under the same laser power.

Globular DNA condensed by other agents such as spermidine and PEG–NaCl can also be trapped. The trapping force is affected by the transformation agent, as shown in Fig. 3.15. These results strongly suggest that condensation of DNA transformed into the globular structure, which may induce a higher refractive index and lower viscosity drag force due to its decreased size, is essential for laser trapping of DNA molecules. Laser trapping

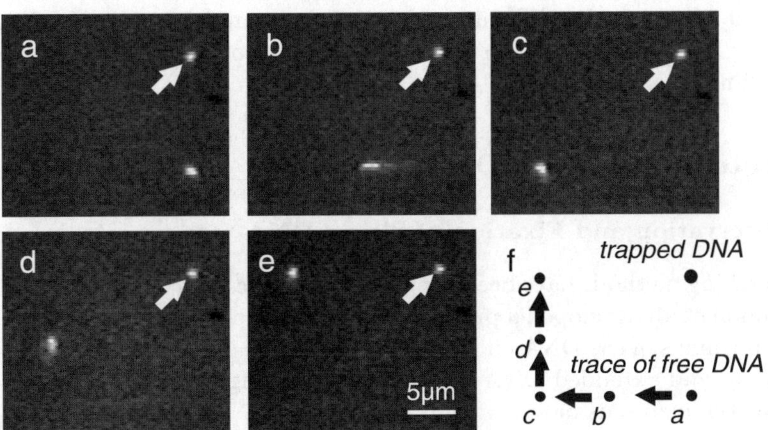

Fig. 3.14. Direct laser trapping of globular T4 DNA. Sequential photographs of trapped and free T4 phage DNA molecules in the globular state at $t = 0$ (**a**), 0.24 s (**b**), 0.64 s (**c**), 0.87 s (**d**), and 1.14 s (**e**). Scale bar: 5 μm. The position of the optical trap is indicated by *white arrows*

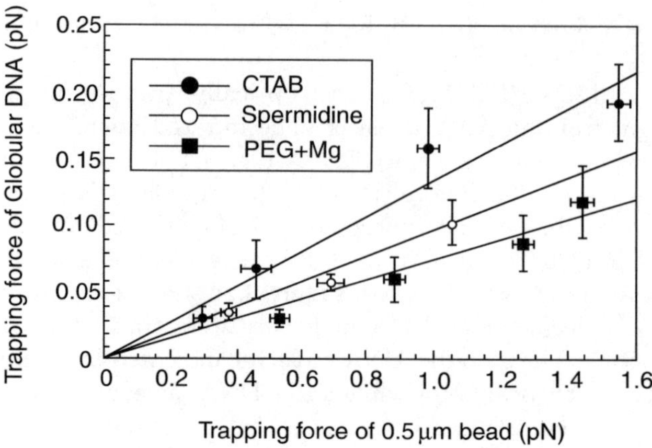

Fig. 3.15. Trapping force on globular DNA using various reagents. The trapping force depends on the condensing reagent. CTAB produced the strongest force

of a single supercoiled lambda DNA using a krypton ion laser operating at 647 nm was also reported by Zare et al. [27]. In their experiment, lambda DNA molecules were stained with YOYO-1 dye. It is observed that YOYO-1 dye induces DNA condensation into the globular state within about 30 min in the absence of any condensing agents. This suggests that the trapping mechanism observed by Zare et al. might be attributed to the YOYO-1-induced condensation, but this is still unknown for the DNA structure. On the other hand, the globular transformation induced by condensing agents has been extensively studied by means of spectroscopy and microscopy, so the morphology and experimental conditions regarding globular DNA are already known [17–19]. It might be difficult to reverse the transformation of DNA condensed by YOYO-1, because YOYO-1 dye binds to the DNA molecule with high affinity.

3.2 Stretching a Giant DNA Molecule

3.2.1 Observation and Fixation of Single DNA

Several stretching methods have been proposed for application of optical mapping. Bensimon et al. developed a process called molecular combing. They anchored one terminus of the DNA on the surface of a glass plate dipped into the DNA solution, and extended DNA molecules by moving the glass plate upward so that the meniscus moved with it, pulling the DNA molecules [28,29]. Yokota et al. also developed a method for straightening DNA molecules for optical restriction mapping. In their approach, DNA molecules in solution at the intersection of a straight-edged cover slip and a slide were also stretched by a moving interface. However, the interface was driven by a moving cover

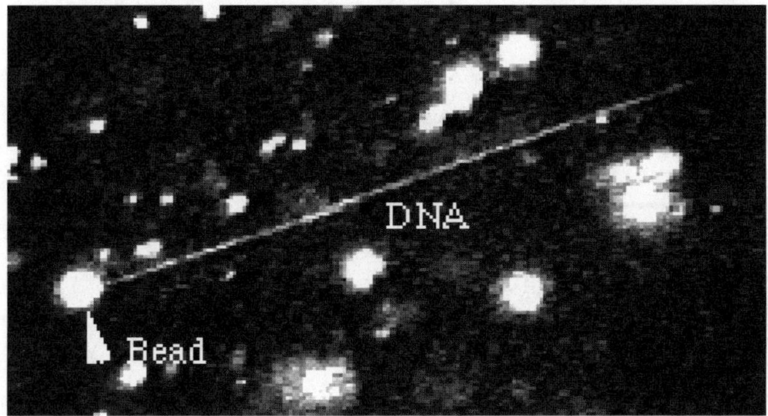

Fig. 3.16. Fixation of a stretched DNA in agarose gel

slip [30]. Despite these developments, it is not so easy to straighten large DNA molecules, because large DNA such as chromosomal DNA is extremely fragile and easily fragmented by shear stress in the accompanying flow of the solution.

DNA molecules can be visualized using a fluorescent dye [1]. We shall now describe an example of the visualization procedure. A T4 DNA (250 μg/ml) solution of 1 μl is mixed with 1 μl of ethidium bromide solution (1 mg/ml in water) for staining. The final concentration of T4 DNA is adjusted to 1 μg/ml by diluting with the TE buffer (10 mM Tris-HCl, 1 mM EDTA, pH 8.0). Fine images of individual DNA molecules can be obtained using an SIT camera and an image processor. DNA molecules frequently change shape due to their Brownian motion. If this motion is suppressed, a more precise image can be obtained. It is possible to suppress Brownian motion by cooling the sample solution [31] or using PEG to increase its viscosity.

Fixation in a gel is another effective method for suppressing Brownian motion [32]. DNA molecules in agarose solution can be stretched by flow forces and then fixed in the stretched shape by gelation brought about by a rapid temperature decrease [33], as shown in Fig. 3.16. The samples in the agarose gel are prepared as follows. The stained DNA is diluted with 1% agarose and loaded on a glass slide. After gelation of the sample, a cover glass is put on and heated to melt the agarose gel once again. The molten agarose solution is spread by the glass cover.

3.2.2 Stretching and Fixing DNA Via the Globule–Coil Transformation

As described previously, the globular transformation can be used to manipulate long DNA molecules. The reversible globular transformation was therefore applied to stretch DNA [34]. First, a single yeast chromosomal DNA

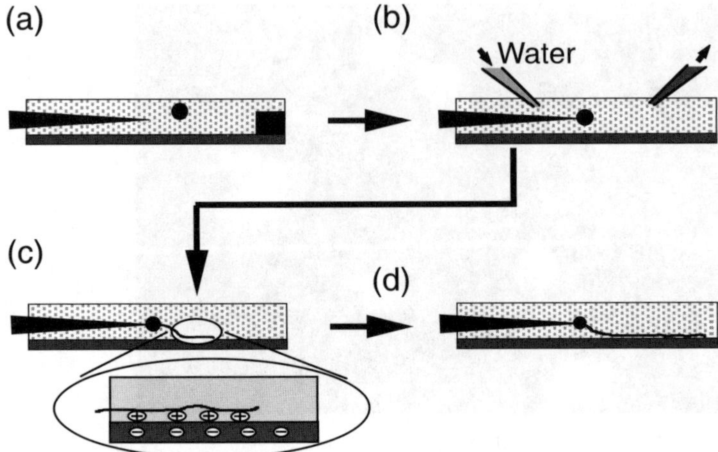

Fig. 3.17. Procedure for sequential extraction of globular DNA. (**a**) Anchor globular DNA on a tip of the needle electrode by gradient force. (**b**) Decondesation by diluting with distilled water. (**c**) Anchor the other end on the glass surface. (**d**) Sequential stretching

molecule was subjected to a globular transformation and the globule was anchored at the tip of a needle electrode made of tungsten. When the globular DNA reverted to the coiled state by reducing concentrations of condensing reagents, the coiled DNA was sequentially spun from the globular DNA like a spindle. By manipulating the tip of the needle electrode, chromosomal DNA was successfully spun and fixed on a glass surface in an arbitrary pattern.

Figure 3.17 is a schematic illustration of the procedure for stretching DNA molecules. As described in the previous section, yeast chromosomal DNA was prepared in agarose gel and transformed to globular structure by soaking in PEG/NaCl solution. The globular DNA molecules were extracted from the gel by diluting the agarose gel. One of the globular DNA molecules was anchored at the tip of a needle electrode (Fig. 3.17a) and transformed to the coiled state (Fig. 3.17b). The coiled DNA molecule was spun from the chromosomal DNA in the globular state and straightened out on the glass surface (Figs. 3.17c and d).

In order to stretch DNA molecules according to the scheme in Fig. 3.17, the condensing reagent was diluted with distilled water. During dilution, the DNA molecule should be anchored to prevent DNA molecules from drifting due to solution flow accompanying buffer exchange. To anchor them, globular DNA molecules were manipulated by a gradient force, because a Coulomb force cannot drive them. Applying 10 V_{eff}, 1 MHz of ac voltage, globular DNA molecules were attracted and attached to the tip of a sharp needle electrode, as shown in Fig. 3.18.

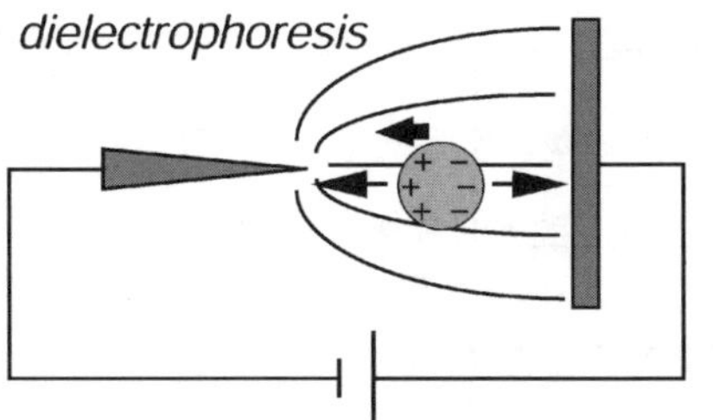

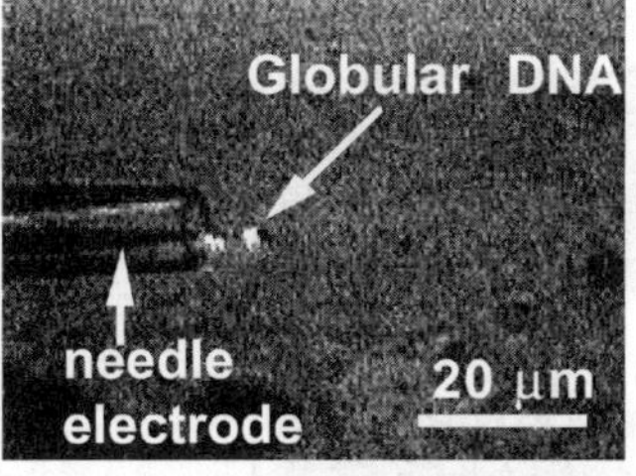

Fig. 3.18. Holding a globular DNA molecule at the tip of a needle electrode by means of a gradient force

After attaching the DNA, the condensing reagent was diluted with distilled water and the globular DNA then unfolded to the coiled state. Because the phase transition to the coiled state progressed from the terminal of the globule, the coiled DNA was sequentially spun from the globular DNA like a spindle. The extracted DNA can be fixed on the surface of a glass cover slip using cross-linked magnesium ions.

Figure 3.19 shows the stretching of a yeast chromosomal DNA molecule. By manipulating the tip of the needle electrode, chromosomal DNA was successfully spun and fixed on a glass surface in an arbitrary pattern. Figure 3.20 is another example of stretching. In this case, the transition was made using ethanol solution. This improved the reliability of the stretching because electrolysis was eliminated due to the decreased conductivity of the solution.

This method has an advantage over the previous methods, because DNA molecules can be stretched to form a pattern, such as a square. This allows one to observe very long DNA molecules in a limited microscope field. From the fluorescence image, no kink was observed during spinning. This result suggests that globular DNA is packed in an orderly way, without entanglement, even in the case of very long molecules such as chromosomal DNA. After spinning out chromosomal DNA, only a very small fluorescence was

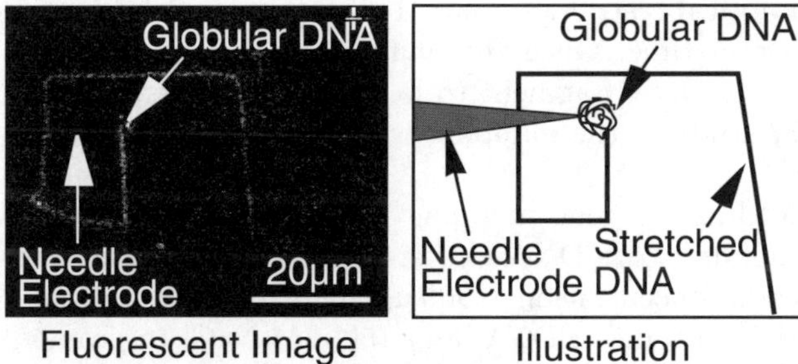

Fig. 3.19. Sequential extraction of yeast chromosomal DNA

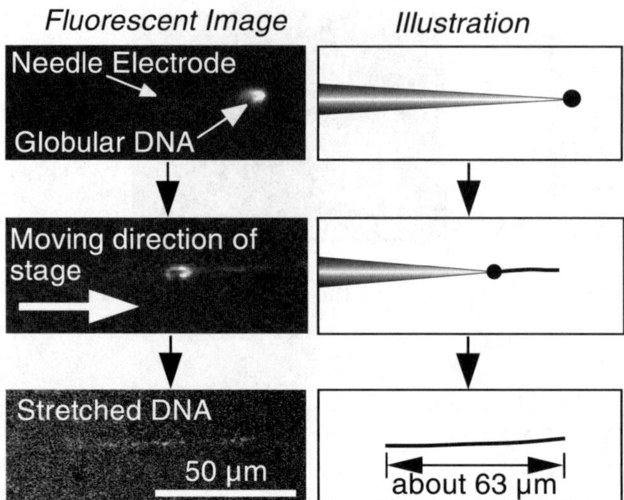

Fig. 3.20. Stretched T4 DNA using the globule–coil transformation in ethanol solution. *Left*: fluorescent image. *Right*: schematic illustration

found to attach to the tip of the electrode. This observation indicates that almost the entire chromosomal DNA in the globular form can be spun and fixed on the glass surface.

3.3 Mapping Stretched Single DNA Molecules

3.3.1 Hybridization with a Probe

Gene mapping is a very important step in enhancing the speed of genome analysis. If a probe DNA of known sequence is hybridized to a denatured single DNA molecule fixed in stretched form, the locations of hybridized points can be observed using a fluorescent dye which only attaches onto the double-strand part of the DNA. A known sequence can therefore be located using an optical microscope. Restriction enzymes with fluorescent dye can also be used to make an optical restriction map. This process can be carried out within a short period of time. Once this mapping has been achieved, the next step is to cut out a DNA fragment to be precisely sequenced. The following is a preliminary study of the mapping technique based on a single DNA molecule.

The preliminary procedure for gene mapping is shown schematically in Fig. 3.21 [35]. Purified lambda phage DNA was first incubated at room temperature for 5 to 10 days at a concentration of 500 μg/ml in a solution containing 10 mM Tris-HCl (pH 8.0), 5 mM EDTA, and 150 mM NaCl to promote annealing between its cohesive ends. It was then diluted with distilled water and reacted with T4 DNA ligase. One molecule of biotin was introduced to

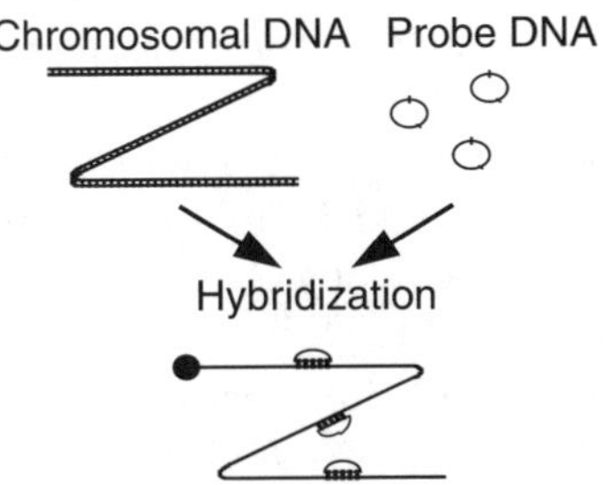

Fig. 3.21. Optical mapping of DNA by probe hybridization

both (l and r) or only one (r) strand of the 3′ end of the ligated DNA, according to Zimmermann and Cox [36]. These molecules (tandem lambda DNA) were used as hybridization templates. The length of the tandem lambda DNA was evaluated by pulsed-field gel electrophoresis. The tandem lambda DNA (1 µg/ml) was denatured in 0.15 M NaOH, neutralized with HCl, and then hybridized with 30 times molar excess (2.6 µg/ml) of a single-stranded DNA probe to the corresponding sequence on the template DNA. The reaction was carried out at 42°C for 16 hr in 0.3 M NaCl, 20 mM HEPPS–HCl (pH 8.3), and 2 mM EDTA. After hybridization, magnetic beads (Dynabeads M-280 streptavidin, DYNAL) were added to a concentration of 1.5 mg/ml and incubated at 30°C for 12 h using gentle rotation. The beads were then washed with 10 mM Tris-HCl, 2 mM EDTA, and 50 mM NaCl using a magnetic particle concentrator (DYNAL MPC-P-12) to remove free template and probe DNAs.

Figure 3.22 shows a microscope image of bead-bound hybridized DNA stretched in agarose gel. Three glowing spots (indicated by arrows), corresponding to the double-stranded region formed by probe hybridization, were observed along a line at regular intervals from the bead. The interval length is

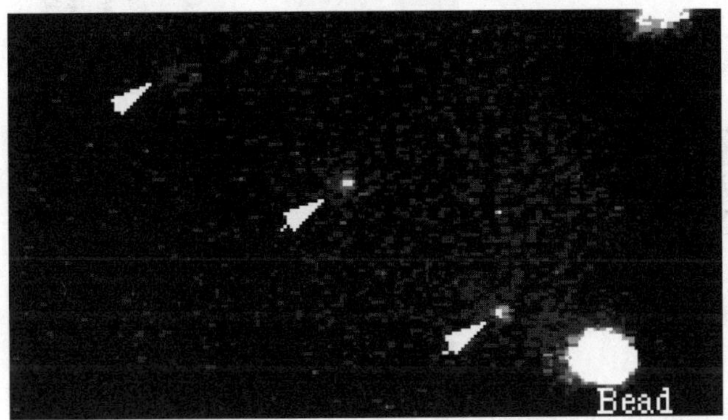

Fig. 3.22. Visualization of a probe hybridized to a stretched DNA

about 20 μm, equal to the length of single-stranded lambda phage monomeric DNA.

This is still a preliminary result. The single-stranded portion of hybridized DNA cannot be observed by this staining condition, and it is not certain whether the single-stranded part of the DNA is straight or not. In agarose gel, the single-stranded part may not be extended, as the DNA is relatively firmly fixed, so that the fluorescent points may represent the structure of the template DNA. However, if the medium is liquid, the single-stranded part may get tangled. To increase accuracy, the process should be carried out under the condition that the DNA is always stretched in order to avoid tangling. For example, single-stranded DNA could be visualized using a single-stranded binding protein (SSB). After these improvements, this gene mapping technique may be applicable to various DNA molecules without substantial modification of the widely used vectors.

3.3.2 Restriction Map

Restriction enzymes attach to a recognized part of double-stranded DNA and this can be used to make an optical restriction map. The localization technique of the enzyme reaction, described in Sect. 3.4, contributes to making the optical restriction map. The restriction enzyme is stained with a fluorescent dye. Staining is carried out on restriction enzymes attached to DNA in a solution without Mg ions. This protects active sites on the enzyme from fluorescent dye attachment. DNA molecules in solution without Mg ions are bound to a magnetic bead and EcoRI is mixed to attach to the DNA. DNA molecules bound to the magnetic bead are condensed using a magnet and

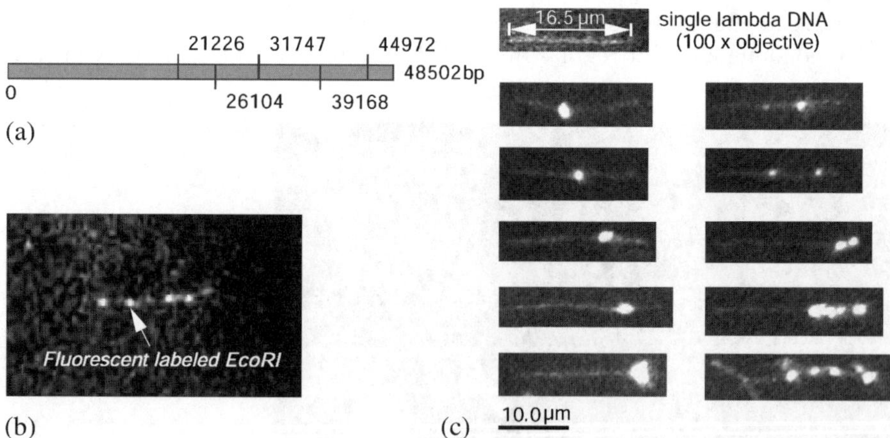

Fig. 3.23. Optical restriction mapping of fluorescent labeled EcoRI on stretched single lambda DNA. (**a**) Five EcoRI restriction sites on lambda DNA. (**b**) Fluorescent labelled EcoRI. (**c**) Single lambda DNA

excess enzymes removed. The amine-reactive dye Oregon Green 488 is then applied for staining and DNAs are again condensed. After this procedure, the stained EcoRI is detached from the DNA by adding $MgCl_2$ solution.

Figure 3.23 shows the optical restriction map of lambda phage DNA with stained EcoRI [37]. Fluorescent-labeled EcoRI bound to five specific sites on stained DNA straightened on the cover slip in the absence of Mg ions. The positions of the binding sites agreed with the restriction map when they were normalized to 21 226 nucleotides in the standard sequence data.

These results indicate the possibility of gene mapping a single DNA molecule. DNA micromanipulation techniques have the advantage that DNA can be analyzed in nearly physiological conditions. It is possible to observe reactions in a liquid environment in real time and to use the observed DNA for the next step of precision analysis. This technique can also be used to analyze various chemical reactions among biological molecules, e.g., the dynamics of transcription or replication mechanisms.

3.4 Cutting Stretched DNA

It is important to develop good cutting methods in order to prepare fragments for subsequent investigation. Several methods have been proposed for cutting single DNA molecules. Although a focused ultraviolet (UV) laser can cut DNA molecules [9,26,38], UV irradiation causes serious damage to DNA molecules, such as the formation of pyrimidine dimers. Cutting methods based on local activation of a restriction enzyme are more suitable for this purpose, because the restriction enzyme does not cause any damage, and the DNA retains a biologically active terminus at the cutting site. Washizu et al. cut DNA molecules using a latex bead coated with a restriction enzyme [39]. The bead was controlled by laser trapping and attached to stretched DNA molecules.

A suitable temperature and a divalent cation, such as the magnesium ion, are required to activate most restriction enzymes. A cutting method based on control of local temperature [40] involves raising the temperature locally by means of a focused infrared laser. Hoyer et al. reported a cutting method based on local release of magnesium ions [41]. Ultraviolet radiation induced UV photolysis of a caged compound of magnesium ion. The locally released magnesium ion activated a restriction enzyme Apa I and cut DNA molecules. Because UV irradiation sometimes damages DNA molecules, a cutting method free from UV irradiation is preferable.

Another method has been developed which is based on control of local magnesium ion concentrations. Magnesium ions are essential for activating the restriction enzyme. Because DNA molecules are suspended in a buffer containing a chelater, ethylenediaminetetraacetic acid (EDTA), the magnesium ion concentration in the suspension is extremely low, and the activity of the restriction enzyme is completely suppressed. The restriction enzyme can be activated locally by raising the magnesium ion concentration in that location. The local magnesium ion concentration can be controlled using an electrochemical method. A needle electrode made of magnesium metal or a capillary containing a gel with Mg ions can be used. A dc voltage can control the release of magnesium ions from the needle or capillary. The balance between release and diffusion of magnesium ions determines the distribution of these ions. DNA molecules can be cut locally by this method.

3.4.1 Localizing Enzyme Activity by Local Temperature Control

A novel method, referred to as a local temperature control technique, uses an infrared laser to increase the temperature in a limited area (about 10 μm diameter) under a fluorescence microscope, as shown in Fig. 3.24. Sample temperatures are maintained below a few degrees C or the sample is frozen to eliminate enzyme activity. Using this method, a single DNA molecule in a frozen sample can be transported without attaching a bead [40]. A DNA molecule can also be cut by activating the enzymatic reaction in a limited area of diameter approximately 20 μm. This technique not only manipulates single molecules, but also serves as a tool for promoting chemical reactions locally within large single molecules. Local temperature control techniques may be applied for sequential cutting of DNA fragments. If the localized area is traversed along a stretched single DNA molecule, fragments can be cut successively from one terminus, preserving order information.

As shown in Fig. 3.24, an inverted fluorescence microscope was modified to introduce the Nd:YAG laser required for the local temperature control technique, and an objective lens was used for imaging DNA molecules and focusing the laser beam. A CCD camera monitored non-fluorescent images and a SIT camera monitored fluorescent images of the DNA molecules. These

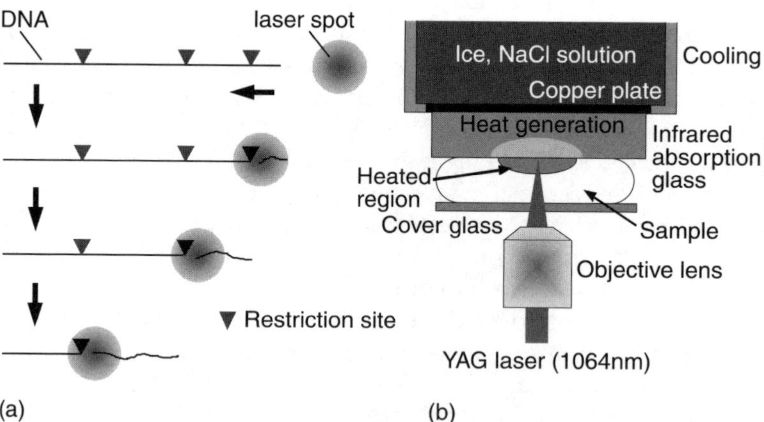

Fig. 3.24. Cutting stretched DNA by local temperature control. (**a**) Schematic diagram. (**b**) Apparatus

images were processed by an image processor and recorded by VCR. The infrared absorption glass for the YAG laser beam was set above the sample. The glass was heated by laser irradiation. Since the sample solution was continuously cooled, adjusting the laser power and/or lowering the temperature could control the temperature of the heated area. In addition, the laser spot could be traversed under microscope view. The cooler used a mixture of solidified carbon dioxide and methanol (approximately −75°C) or ice–NaCl mixture (approximately −2°C) as a cooling medium.

Yeast (Saccharomyces cerevisiae) chromosomal DNA was embedded in agarose gel to suppress Brownian motion and fix the stretched shape. This DNA was visualized by a fluorescent dye, YOYO-1 iodide, excited by incident light at wavelength 491 nm and emitting light at 509 nm. To prevent photobleaching, the DNA solution was supplemented with 1% (v/v) 2-mercaptoethanol, 18 μg/ml catalase, 2.3 mg/ml D−(+)-glucose, and 0.1 mg/ml glucose oxidase. Otherwise the fluorescent dye was photobleached within several tens of seconds. Using the local temperature control technique, the temperature was raised at the target position and the DNA molecules were digested by the restriction enzyme only in the temperature-raised area.

Samples were prepared as follows. A gel plug containing yeast chromosomal DNA was melted and the melted gel was loaded on a cover slip. The cover slip was cooled to 4°C for several minutes. The DNA molecules were fixed and extended in agarose gel. The DNA sample was mounted on a microscopic stage and then the whole cover slip was cooled to approximately −2°C, but not frozen. This was enough to suppress restriction enzyme activity. A reaction buffer (50 mM Tris-HCl at pH 7.5, 10 mM MgCl, 100 mM NaCl, 1 mM dithiothreitol) containing the restriction enzyme EcoRI was diffused from the edge of the cover slip containing the DNA molecules fixed in the gel. When the laser irradiated the cover slip, single DNA molecules in the gel were cut with EcoRI by raising the temperature in the local area just around the laser spot.

To confirm local temperature control, the size of the melted area in the frozen solution was measured. As shown in Fig. 3.25, local area melting was achieved when the solution temperature was −35°C and the laser power

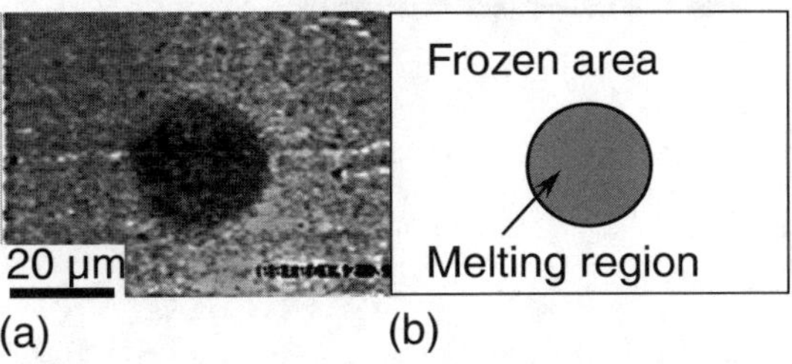

Fig. 3.25. Melting region under the local temperature control method. (**a**) CCD camera image. (**b**) Schematic diagram

was 10 mW. An area of diameter about 20 μm around the laser spot was melted. The image in the figure was obtained by a CCD camera with contrast enhancement. The diameter of the melted area depended on the solution temperature and the laser power. In the laser power range from 7–13 mW, the diameter of the melting area increased gradually with increasing laser power.

Inhibition of restriction enzyme activity due to the fluorescence dye was checked, as was the dependence on Mg ion concentration and temperature. Although dyes are needed to visualize DNA molecules, these may inhibit restriction enzyme activity, depending on their type and concentration [42]. Furthermore, magnesium ions reduce the contrast in fluorescence imaging. Figure 3.26 shows the results of gel electrophoresis to check the effect of fluorescent dye YOYO-1 and Mg ions. Figures 3.26a, b, and c show concentration ratios DNA [nucleotides]:dye [molecules] of 20:1, 10:1, and 1:1, respectively. The Mg ion concentration ranged from 0 to 10 mM. Results show that YOYO-1 inhibits restriction enzyme activity at high dye concentrations. Dyes did not affect activity at ratios less than 20:1. The restriction enzyme is activated by more than 0.5 mM Mg ions at the dye ratio 20:1.

Temperature dependence was also checked. Figure 3.27 shows the temperature dependence of restriction enzyme activity. Activity of the enzyme, EcoRI, was suppressed at less than 4°C.

Figure 3.28a shows a bundle of chromosomal DNA molecules fixed in agarose gel. In this procedure, the sample temperature was maintained at 0°C. The DNA bundle was cut by the restriction enzyme EcoRI, activated by the temperature increase in the laser spot (YAG laser power 100 mW). Laser irradiation alone did not cut the DNA. The size of the activated region was about 20 μm in this experiment. The size can be measured by observing the region where the DNA has been cut. A single DNA molecule can also be

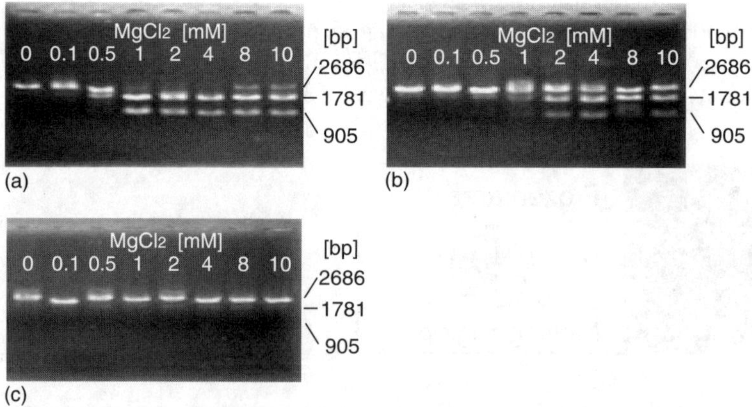

Fig. 3.26. Dependence of EcoRI activity on YOYO-1 and $MgCl_2$ concentrations. (**a**) DNA : YOYO = 20 : 1, (**b**) DNA : YOYO = 10 : 1, (**c**) DNA : YOYO = 1 : 1

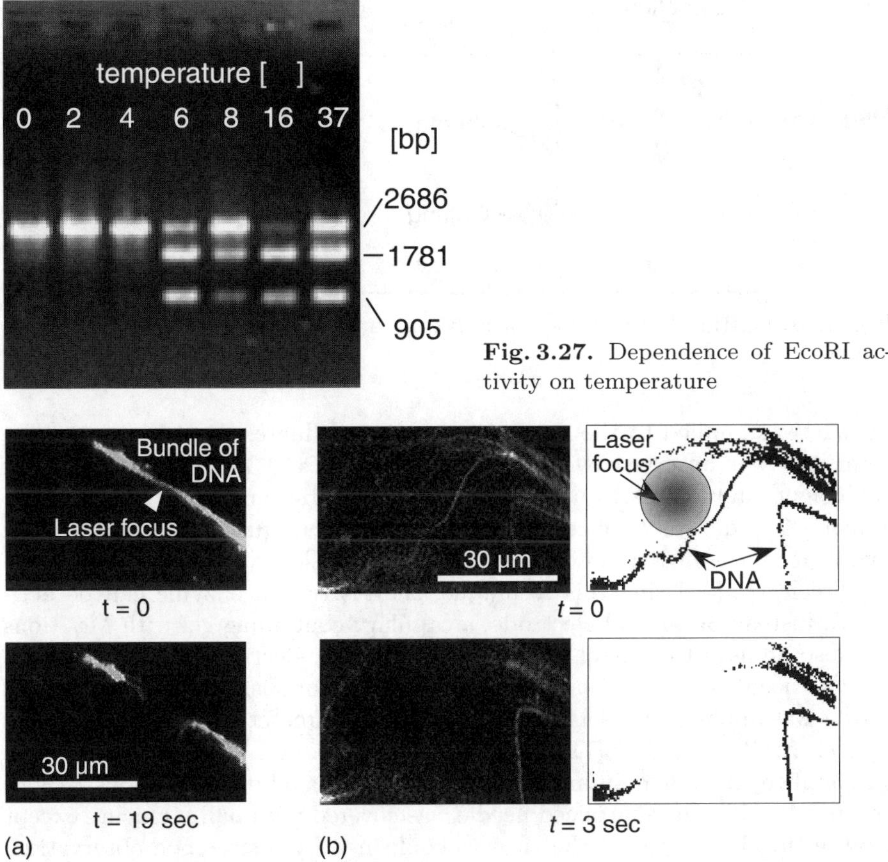

Fig. 3.27. Dependence of EcoRI activity on temperature

Fig. 3.28. Cutting a bundle of DNA molecules by local temperature control. (**a**) Bundle of yeast chromosomal DNA. (**b**) Single yeast chromosomal DNA

cut in the same manner (Fig. 3.28b). In this method, the resolution of the cutting position is not very accurate. Successive fragments can be obtained by moving the laser along from one terminus. Using restriction enzymes which recognize 8 base pairs or more, the distance between adjacent restriction sites becomes larger, so that a single restriction site could be cut independently by this method.

3.4.2 Cutting DNA by Controlling Ionic Concentration

Control of local Mg ion concentration is another useful method for cutting stretched DNA molecules, as shown in Fig. 3.29 [43]. To demonstrate such control, the distribution of magnesium ions was observed in a buffer containing 1 mM EDTA, 1 mM dithiothreitol (DTT) and 0.1 μM magnesium green, a fluorescence dye that specifically binds to the magnesium ion. When a dc

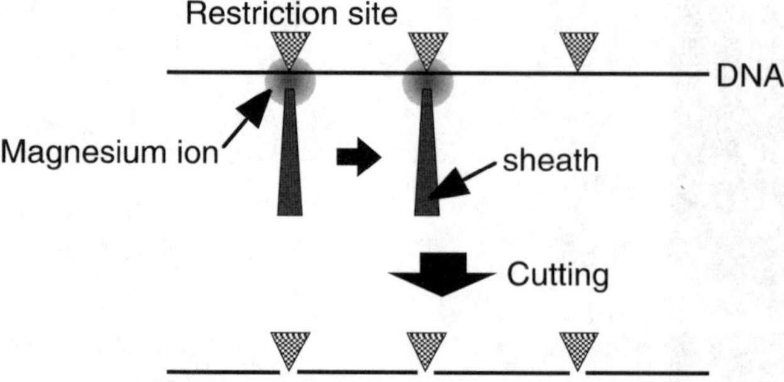

Fig. 3.29. Cutting DNA by locally activating a restriction enzyme under Mg ion control

voltage was applied to the needle electrode, the fluorescence due to magnesium ions was observed by the microscope. Figure 3.30 shows an image of the magnesium ion distribution visualized by magnesium green fluorescence. When +5 V dc was applied to the needle electrode made from magnesium metal, magnesium ions were electrochemically released from the tip of the needle electrode. Using this technique, the restriction enzyme can be activated. Instead of a metal electrode, a capillary containing gel with Mg^- ions can also be used to control the release of Mg ions electrically.

The needle electrode was fabricated by electropolishing. A dc voltage of +10 V was applied across a magnesium wire of diameter 1 mm and a stainless steel electrode in 0.5 M EDTA solution (pH 8.0) for 60 s. While this dc voltage was applied, the magnesium wire was vibrated up and down at about 1 Hz to sharpen its tip. The sharpened needle was covered with nail manicure except for the tip. The shape of the tip was confirmed by microscope observation. The needle electrode was mounted on the stage of the microscope.

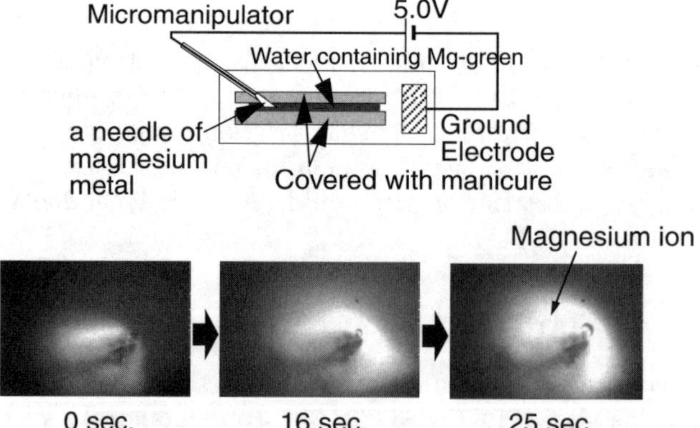

Fig. 3.30. Controlling magnesium ion release by the electrochemical method

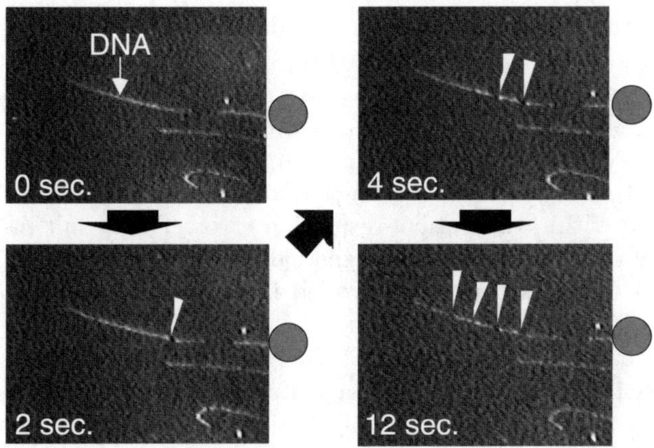

Fig. 3.31. Sequential photographs of DNA cutting by Mg ions. *Filled circles*: tip of needle electrode. *White triangles*: site of cutting by restriction enzyme

Figure 3.31 shows sequential photographs of DNA digestion using this method. The DNA was stretched and fixed, and Mg ions were removed to begin with. Then EcoRI was introduced to bind to the DNA. The sample was set under a microscope and cut. Mg ions were supplied by applying a dc voltage to the magnesium metal needle electrode. During the application of

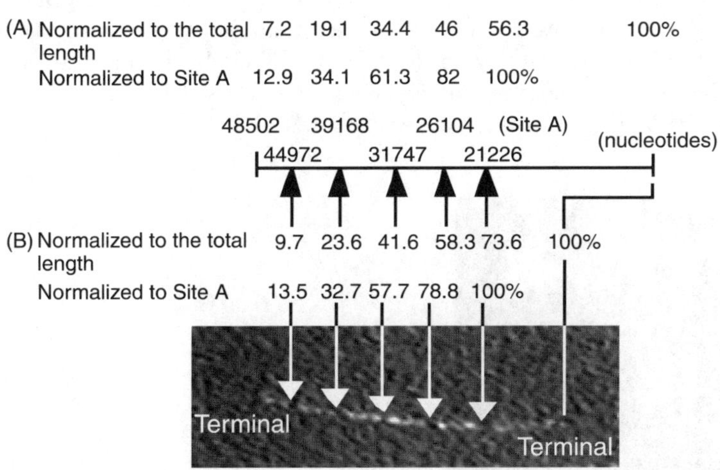

Fig. 3.32. (**a**) Comparison between the restriction map and the optical mapping of EcoRI recognition sites. The restriction map was obtained from lambda DNA sequence data. Each site was normalized to the total length (*upper line*) and to site A (*lower line*). (**b**) Magnified image of cut DNA. The position of each cut site was determined and normalized to the total length (*upper line*) and to site A (*lower line*)

−2 V dc voltage, no cutting reaction was observed (data not shown). However, after applying a +2 V dc voltage, the cutting reaction was initiated. DNA molecules were observed to be cut only in a restricted area close to the tip, and no cutting was observed in areas further away from the tip. The image of the cut DNA was magnified and compared with the EcoRI restriction map obtained from the lambda phage DNA sequence, as shown in Fig. 3.32. They agreed within the optical resolution error. This result indicates that the activity of the restriction enzyme can be controlled by Mg ions and that release of Mg ions can be controlled by dc voltage application.

It should be noted that, if a stretched DNA attaches too strongly to a substrate, the hybridization or attachment of a restriction enzyme will be inhibited. The reliability of the method can only be improved by adjusting the attachment condition and carefully evaluating this inhibition.

3.5 Recovery of DNA Fragments

DNA fragments can be recovered in a glass capillary (internal diameter 18 μm) using electrophoresis, as shown in Fig. 3.33. A positive dc voltage was applied to the capillary. Being negatively charged, DNA fragments were transported into the capillary by the dc electric field [25].

Figure 3.34 shows a microflow system for separating DNA fragments. The DNA molecule is fixed at port G and cut [44]. Ports A, B and C are supplied with a dc voltage to drive the fragments by electrical force. Switching the applied voltage, the fragments are separated. Figure 3.35 shows the recovery of a fragment. In this case cutting was achieved by a pulsed UV laser. The dc voltage was +20 V and fragments were recovered at ports A or B.

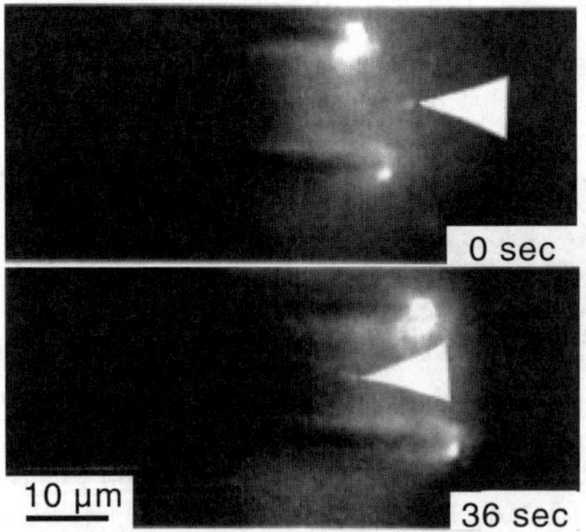

Fig. 3.33. Recovering a DNA fragment in a capillary by electrophoresis

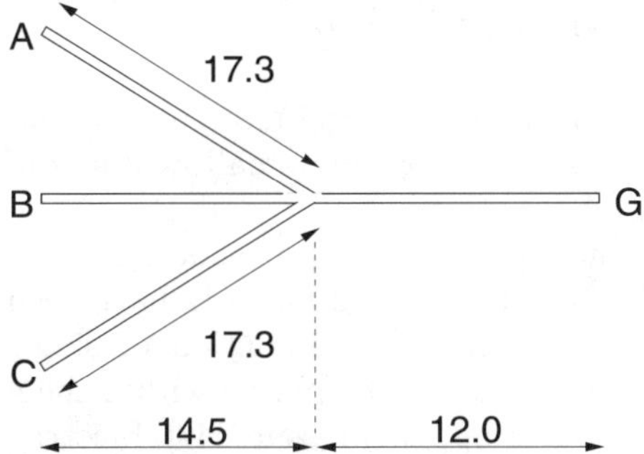

Fig. 3.34. Geometry of the selective recovery device for single DNA manipulation. The width and depth of channels were 800 μm and 500 μm, respectively. The electrode G was earthed, whilst electrodes A, B, and C were used to apply a positive potential (+20 V). Dimensions are given in millimeters

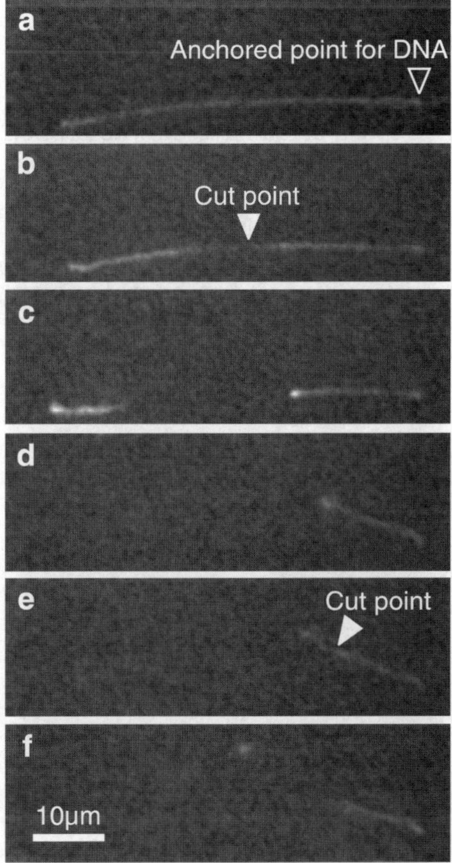

Fig. 3.35. Sequential photographs showing the continuous cutting and recovery of single DNA molecules

3.6 Microreactors for DNA Manipulation

The conventional test tube is too large for handling single molecules. A microreactor, or liquid droplets, serve as containers for very small amounts of molecules and agents.

To establish a microreactor, the contents of the reactor should be isolated from its exterior. Liposomes, vesicular phospholipid bilayers, have been proposed as a microreactor and used for indirect manipulation of single molecules, because single water-soluble molecules can be placed within them and then manipulated by electrophoresis [45] or optical pressure [46]. Electric-field pulses can achieve electrofusion of liposomes, which in turn induce chemical reactions between two molecules from two different liposomes [46]. However, the probability of proper fusion between two liposomes is limited and repeated application of the pulses generates leakage from the containers. Furthermore, microreactors based on liposomes cannot be applied to PCR (polymerase chain reaction) amplification, because they are unstable under high temperature conditions [47].

The interface between oil and water works as a barrier for both hydrophobic and hydrophilic molecules. An individual oil droplet containing zinc tetraphenylporphyrin was optically trapped and analyzed by absorption microspectroscopy [48]. This property is also suitable for localizing biological molecules such as enzymes and their substrates. The water-in-oil emulsion has therefore been applied in certain fields of biological research such as enzymology [49] or molecular evolution [50].

A microreactor system using water droplets in oil has been developed as shown in Fig. 3.36. Because water-soluble molecules in the water droplet are kept inside the droplet, the droplet functions as a microreactor. In this text, the system is referred to as a W/O microreactor system and the water

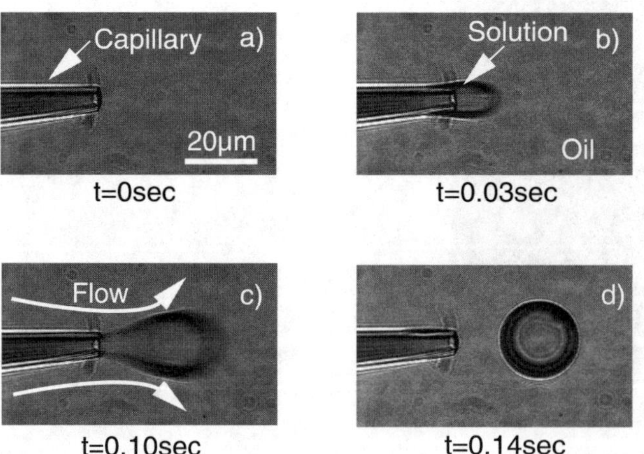

Fig. 3.36. Formation of a liquid droplet in oil for the microreactor system

droplets are referred to as microreactors. The microreactor is easily observed and manipulated under a microscope. This system provides an efficient way of handling single DNA molecules. The fusion between different microreactors containing a single substrate molecule and an enzyme, respectively, can control an enzymatic reaction for a single molecule. The microreactor requires extremely small quantities of sample. This microreactor system promotes a rapid chemical reaction, because the reaction volume is extremely small and the time required for diffusion in the microreactor is thus very short [51]. It should be noted that the temperature of this system can be controlled over a wide range and PCR amplification can be carried out.

3.6.1 Production of Microreactors in Oil

The sample solution in a glass capillary was extruded into rapeseed oil (Nacalai Tesque) by a micro-injector (NARISHIGE, IM-300), and the microreactors were generated. The capillary was mounted on a micromanipulator (NARISHIGE, MHW-13) for precise position-control of the tip. The glass capillary was prepared by NARISHIGE PB-7 and its tip sharpened to a diameter of about 10 μm in diameter. The sample solution contained yeast chromosomal DNA or fluorescent dye, and 0.5% of Tween 20, a non-ionic surfactant, was added to reduce the surface tension and thereby stabilize the microreactors [52].

3.6.2 Manipulation and Fusion of Microreactors

The microreactor manipulation technique can be applied not only to transport water-soluble molecules but also to control chemical reactions between different molecules by fusing two microreactors.

Optical force is also useful for non-contact micromanipulation, and it has already been applied to manipulate cells or biological molecules such as DNA. Microreactors can be manipulated by such techniques. However, in contrast to latex beads or globular DNA molecules, a laser beam repels the microreactors in oil, because water has a smaller refractive index (1.33) than rapeseed oil (1.47). We have also applied this repulsive force for transportation and fusion of microreactors.

The laser is focused in the immediate vicinity of the microreactor to be manipulated. The microreactor is then driven by the repulsive force due to the focused laser beam (Fig. 3.37a). When the microreactor is driven up against another one, fusion is induced (Figs. 3.37b–c). Repeating this operation, we succeeded in fusing four microreactors in 87 s (Figs. 3.37d–g). During the manipulation, vigorous fluctuations of the droplet were not observed. We may attribute this to the fact that the high viscosity of oil prevents Brownian motion. Suppression of Brownian motion indicates that the W/O emulsion system is suitable for manipulating microreactors.

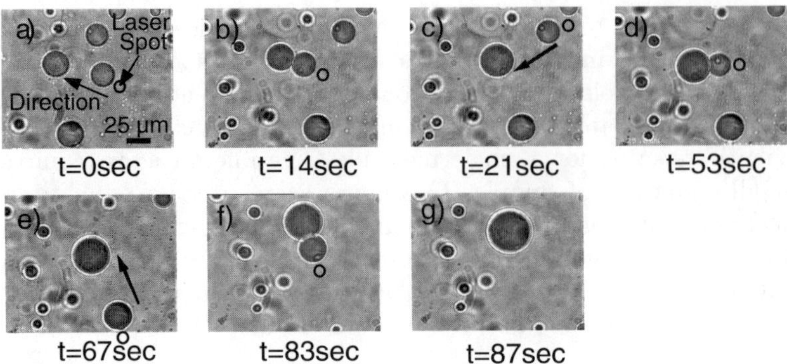

Fig. 3.37. Fusion of microreactors by optical pressure from a laser

3.6.3 Indirect Manipulation of Globular DNA Molecules

Figure 3.38 shows an indirect manipulation of yeast chromosomal DNA. The DNA molecules were transformed to condensed globular structure. The globular DNA molecules were stained with a fluorescent dye, 0.1 µM YOYO, while 1% v/v 2-mercaptoethanol, 0.18 µg/mL catalase, 2.3 mg/mL D−(+)-glucose and 100 µg/mL glucose oxidase were added to suppress photobleaching. The globular transformation was induced by 6% of polyethylene glycol (mol. wt. 6 000) and 0.6 M NaCl. The concentration of yeast chromosomal DNA was adjusted to zero or one DNA molecule in one microreactor, and two microreactors each containing a single globular DNA were selected for the experiment ($t = 0$). The upper two images in Fig. 3.38 were obtained in different focal planes: the two globular DNA under investigation could not be visualized

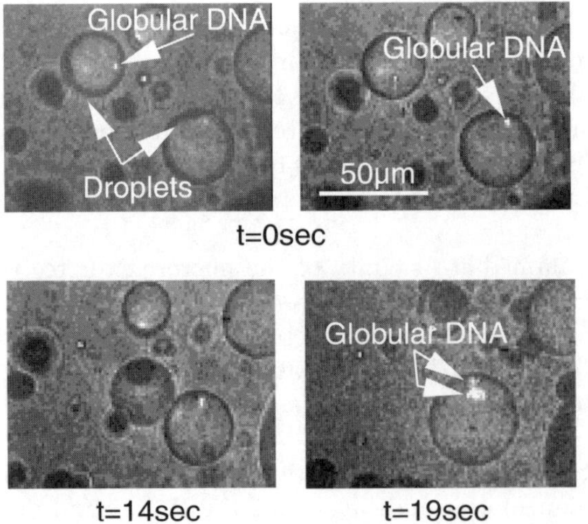

Fig. 3.38. Indirect manipulation of globular DNA by a microreactor

simultaneously due to the small range of fine focus. Concerning the image at $t = 14$ s, the globular DNA molecules were retained within the microreactors whilst the latter were manipulated. Contact between the two microreactors induced them to fuse together and the two globular DNA molecules were mixed inside the fused reactor (Fig. 3.38, $t = 19$ s).

3.6.4 Chemical Reaction in the W/O Microreactor System

These microreactors can be manipulated by optical force. When used to control the fusion between two selected microreactors, this technique can be made to promote chemical or enzymatic reactions locally within the microreactors. The technique can also be used to design a miniaturized total analysis system (μ-TAS).

As a preliminary experiment, we carried out binding of DNA and YOYO. YOYO is a fluorescent dye with an extremely low fluorescence signal when not bound with DNA, but which produces a fluorescent signal when bound [53]. Figure 3.39 shows a microscope image to demonstrate binding of DNA and YOYO inside the microreactors. The experimental process was as follows. First, two kinds of microreactor were produced, one containing yeast chromosomal DNA and the other containing 1 μM YOYO and 1% v/v 2-mercaptoethanol, 0.18 μg/mL catalase, 2.3 mg/mL D−(+)-glucose and 100 μg/mL glucose oxidase to suppress photobleaching. Both reactors were supplemented with Tween 20 to a final concentration of 0.05% to control surface tension. The microreactors containing YOYO were easily distinguished under irradiation by their stronger excitation light, because even YOYO free

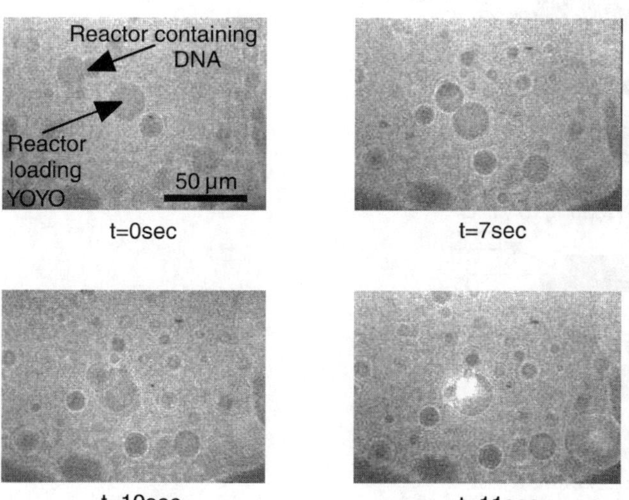

Fig. 3.39. Chemical reaction between DNA and a dye using the microreactor system

from DNA has a weak fluorescence signal. These microreactors were put on a cover glass and manipulated by the optical force of the laser. When a fusion was induced between microreactors, the YOYO bound immediately with DNA (within 0.5 s) and emitted a high fluorescence signal.

3.6.5 PCR Amplification of DNA Fragments

DNA amplification by PCR (polymerase chain reaction) is an important tool for the single-molecule technique, because conventional analytical apparatus cannot handle single molecules. However, PCR amplification requires known DNA sequences at both ends, and therefore known DNA fragments, called linker DNA, must be connected for PCR amplification of unknown DNA fragments. The connection procedure in Fig. 3.40 was developed to improve the connection efficiency of this linker DNA. The linker DNA was designed to be compatible with the ends of the target DNA and consisted of two oligonucleotides, 12 nt and 30 nt. DNA ligase connected the larger oligonucleotide to the target DNA. However, the smaller oligonucleotide was not connected with the target DNA due to the lack of a phosphorous group at the 5′ terminus. The smaller oligonucleotide was easily detached by raising the temperature. The single-strand region can then be synthesized by a heat resistive polymerase such as Taq DNA polymerase. The DNA fragments thus produced can be amplified by a standard PCR procedure. Using this procedure, 100 molecules of a 517 bp fragment were successfully amplified [54]. Because PCR

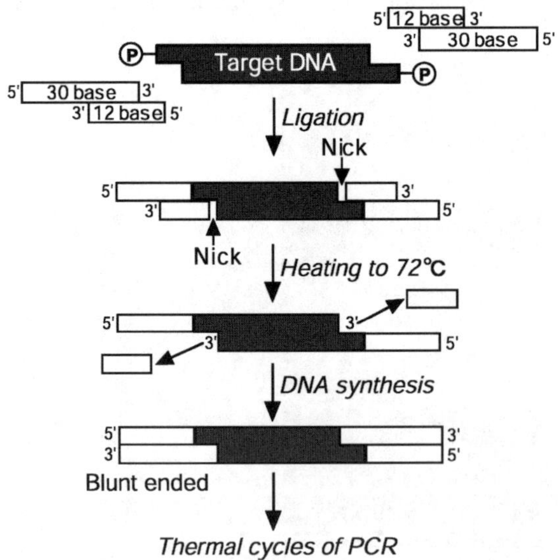

Fig. 3.40. Procedure for amplifying an unknown DNA fragment by connecting linker DNA

amplification of a single DNA molecule is quite difficult, PCR must be further improved in order to amplify a single unknown DNA fragment.

A microreactor system with liquid droplets in silicone oil was used to conduct the PCR, as shown in Fig. 3.41. The concentration of DNA fragments was reduced to about 1 molecule on average in 50 μl of the sample solution with PCR agents. This sample solution was mixed with 0.95 ml silicone oil and liquid droplets were made in the silicone oil. Under these conditions, one droplet contained at most one DNA fragment to be amplified, and most of the droplets did not contain any fragments. Using this water/oil emulsion, the PCR thermal cycle was repeated 13 times. Then the emulsion was centrifuged to separate the liquid and silicone oil. After this process, the thermal cycle was repeated 25 times. Figure 3.42 shows electrophoresis results for the PCR products obtained from a single DNA fragment using the procedure described above.

These experimental results bring out the advantages of using a fine droplet as a container for a single molecule. They show that the loss of a single molecule can be avoided and that the effective concentration of the sample can be increased through the reduced volume. It should also be stressed that this micro-droplet system is suitable for a microchannel chemical system to be constructed on a silicon surface.

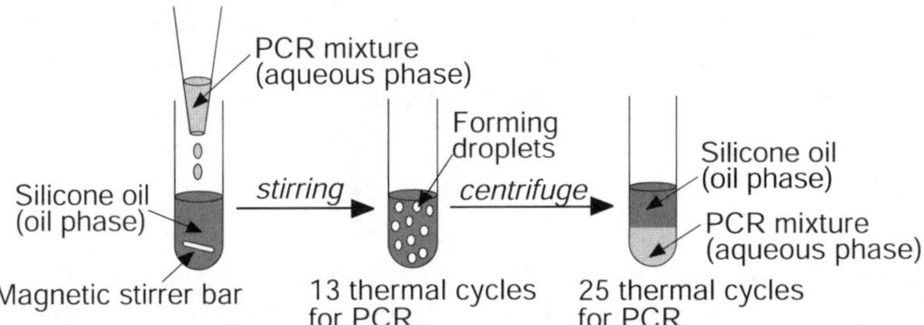

Fig. 3.41. Experimental procedure for PCR of a single DNA fragment using the W/O emulsion

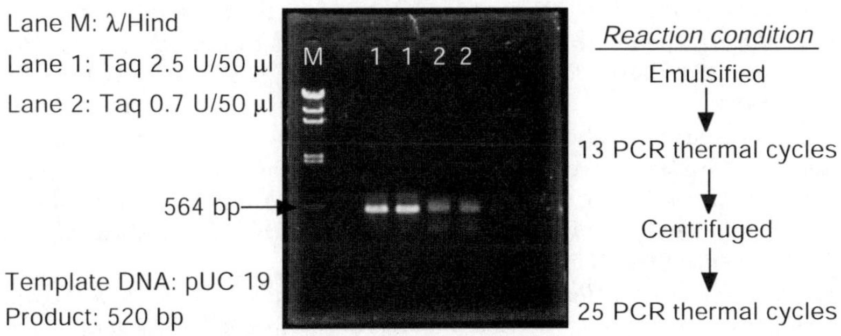

Fig. 3.42. PCR products from the W/O emulsion

3.7 Conclusion

Micromanipulation of single DNA molecules provides a novel analytical method in biotechnology. Laser trapping and electrostatic force are useful tools for manipulating single DNA molecules. The following manipulation techniques have been introduced:

- manipulation of large chromosomal DNA using the globular transformation to suppress damage due to shear stress;
- stretching of single DNA molecules using the reverse transformation to the coiled state;
- optical mapping by probe hybridization or restriction enzymes;
- cutting of a stretched single DNA by raising the temperature locally or releasing Mg^- ions from a tip to control activity of the restriction enzyme;
- the microreactor system to conduct chemical reactions based on single molecules.

These manipulation techniques can be incorporated into an integrated microfluidic device. As pointed out, there are many advantages in using a method which can conduct chemical reactions with reduced sample volume. The most important is that the speed of chemical reactions can be improved because molecules have a shorter distance to diffuse in order to react. Further improvement of these manipulations will enhance the speed of DNA sequencing and contribute to our understanding of interactions, such as DNA/protein interactions, between individual biological molecules.

References

1. K. Morikawa, M. Yanagida: *Visualization of Individual DNA Molecules in Solution by Light Microscopy: DAPI Staining Method*, J. Biochem. **89**, 693–696 (1981)
2. I. Sase, H. Miyata, J.E.T. Corrie, J.S. Craik, K. Kinoshita Jr: *Real Time Imaging of Single Fluorophores on Moving Actin with an Epifluorescence Microscope*, Biophys. J. **69**, 323–328 (1995)
3. S. Gurrieri, S.B. Smith, K.S. Wells, I.D. Johnson, C. Bustamente: *Real-Time Imaging of the Reorientation Mechanisms of YOYO-Labelled DNA Molecules During 90 and 120 Pulsed-Field Gel Electrophoresis*, Nucleic Acids Res. **24** 4759–4767 (1993)
4. H. Yin, M.D. Wang, K. Svoboda, R. Landick, S.M. Block, J. Gelles: *Transcription against an Applied Force*, Science **270**, 1653–1657 (1995)
5. H. Kabata, O. Kurokawa, I. Arai, M. Washizu, S.A. Margarson, R.E. Glass, N. Shimamoto: *Visualization of Single Molecules of RNA Polymerase Sliding along DNA*, Science **262**, 1561–1563 (1993)
6. D.C. Schwartz, X. Li, L.I. Hernandez, S.P. Ramnarain, E.J. Huff, Y.-K. Wang: *Ordered Restriction Maps of Saccharomyces Cerevisiae Chromosomes Constructed by Optical Mapping*, Science **262**, 110–114 (1993)

7. M. Nishioka, S. Katsura, A. Mizuno: *Manipulation of a Single DNA Molecule Inside a Capillary*, Trans. IEE of Japan **166** E 171–177 (1996) in Japanese
8. T.T. Perkins, S.R. Quake, D.E. Smith, S. Chu: *Relaxation of a Single DNA Molecule Observed by Optical Microscopy*, Science **264**, 822–826 (1994)
9. A. Mizuno, M. Nishioka, T. Tanizoe, S. Katsura: *Handling of a Single DNA Molecule Using Electric Field and Laser Beam*, IEEE Trans. Ind. Appl. **31**, 1452–1457 (1995)
10. A. Ashkin, J.M. Dziedzic, J.E. Bjorkholm, S. Chu: *Observation of a Single-Beam Gradient Force Trap for Dielectric Particles*, Opt. Lett. **11**, 280–290 (1986)
11. A. Ashkin, J.M. Dziedzic, T. Yamane: *Optical Manipulation of Single Cells Using Infrared Laser Beam*, Science **330**, 769–771 (1987)
12. A. Ashkin: *Focus of a Single-Beam Gradient Laser Trap on a Dielectric Sphere in the Ray Optics Regime*, Biophys. J. **61**, 469–582 (1992)
13. M.W. Berns, W.H. Wright, B.J. Tromberg, G.A. Profeta, J.J. Andrews, R.J. Walter: *Use of a Laser-Induced Optical Force Trap to Study Chromosomal Movement on the Mitotic Spindle*, Proc. Natl. Acad. Sci. USA **86**, 4539–4543 (1989)
14. T. Nishizaka, H. Miyata, H. Yoshikawa, S. Ishiwata, K. Kinosita Jr: *Unbinding Force of a Single Motor Molecule of Muscle Measured Using Optical Tweezers*, Nature **377**, 251–254 (1995)
15. M. Nishioka, S. Katsura, K. Hirano, A. Mizuno: *Evaluation of Cell Characteristics by Stepwise Orientational Rotation Using Opto-electrostatic Micromanipulation*, IEEE Trans. Ind. Appl. **33**, 1381–1388 (1997)
16. H.A. Phol: *Dielectrophoresis* (Cambridge University Press, New York 1978)
17. A. Mizuno, M. Washizu: *Biomedical Engineering*, in *Handbook of Electrostatic Processes* (Marcel Dekker, 1995) pp. 653–686
18. R. Huey, S.C. More: *Condensed State of Nucleic Acids III, $\Psi(+)$ and $\Psi(-)$ Conformational Transitions of DNA Induced by Ethanol and Salt*, Biopolymers **20**, 2533–2552 (1981)
19. V.A. Bloomfield: *Condensation of DNA by Multivalent Cations: Considerations on Mechanism*, Biopolymers **31**, 1471–1481 (1991)
20. L.S. Lerman: *A Transition to a Compact Form of DNA in Polymer Solutions*, Proc. Natl. Acad. Sci. U.S.A. **68**, 1886–1890 (1971)
21. C.F. Jordan, L.S. Lerman, J.H. Venable: *Structure and Circular Dichroism of DNA in Concentrated Polymer Solutions*, Nature **236**, 67–70 (1972)
22. T. Maniatis, J.H.J. Venable, L.S. Lerman: *Structure of Ψ DNA*, J. Mol. Biol. **84**, 37–64 (1974)
23. K. Yoshikawa, Y. Matsuzawa: *Nucleation and Growth in Single DNA Molecules*, J. Am. Chem. Sci. **118**, 929–930 (1996)
24. S. Katsura, A. Yamaguchi, K. Hirano, Y. Matsuzawa, A. Mizuno: *Manipulation of Globular DNA Molecules for Sizing and Separation*, Electrophoresis **21**, 171–175 (2000)
25. S. Katsura, K. Hirano, Y. Matsuzawa, K. Yoshikawa, A. Mizuno: *Direct Laser Trapping of Single DNA Molecules in the Globular State*, Nucleic Acids Res. **26**, 4943–4945 (1998)
26. M. Nishioka, T. Tanizoe, S. Katsura, A. Mizuno: *Micromanipulation of Cells and DNA Molecules*, J. of Electrostat. **35**, 83–92 (1995)
27. D.T. Chiu, R.N. Zare: *Biased Diffusion, Optical Trapping, and Manipulation of Single Molecules in Solution*, J. Am. Chem. Soc. **118**, 6512–6513 (1996)

28. A. Bensimon, A. Simon, A. Chiffaudel, V. Croquette, F. Heslot, D. Bensimon: *Alignment and Sensitive Detection of DNA by a Moving Interface*, Science **265**, 2096–2098 (1994)
29. D. Bensimon, A.J. Simon, V. Croquette, A. Bensimon: *Stretching DNA with a Receding Meniscus: Experiments and Models*, Phys. Rev. Lett. **74**, 4754–4757 (1995)
30. H. Yokota, F. Johnson, H. Lu, R.M. Robinson, A.M. Belu, M.D. Garrison, B.D. Ratner, B.J. Trask, D.L. Miller: *A New method for Straightening DNA Molecules for Optical Restriction Mapping*, Nucleic Acids Research **25**, 1064–1070 (1997)
31. M. Nishioka, Y. Sakaguchi, A. Mizuno: *Manipulation of DNA Molecules at Low Temperature*, Proc. of 1993 Annual Meeting of Inst. Electrostatics Japan (1993) pp. 217–220
32. B. Smith, P.K. Aldridge, J.B. Callis: *Observation of Individual DNA Molecules Undergoing Gel Electrophoresis*, Science **243**, 203–206 (1989)
33. K. Hirano, R. Ishii, S. Matsuura, S. Katsura, A. Mizuno: *Novel DNA Manipulation Based on Local Temperature Control: Transportation and Scission*, in *Micro-Total Analysis Systems 1998*, ed. by D.J. Harrison and A. van den Berg (Kluwer Academic Publishers, Dordrecht 1998) pp. 415–418
34. S. Katsua, N. Harada, T. Sato, K. Imayou, Y. Maeda, S. Matsuura, K. Hirano, K. Takashima, Y. Matsuzawa, A. Mizuno: *Stretching and Cutting of a Single DNA Molecule*, 1st Annual International IEEE–EMBS Special Topic Conference on Microtechnologies in Medicine & Biology (2000) pp. 53–57
35. Y. Matsui, K. Takashima, S. Katsura, A. Mizuno: *Development of Gene Mapping for Single Chromosomal DNA Molecules*, Trans. IEE Japan **119** E, No. 12, 620–625 (1999)
36. R.M. Zimmermann, E.C. Cox: *DNA Stretching on Functionalized Gold Surfaces*, Nucleic Acids Res. **22**, 492–497 (1994)
37. S. Matsuura, K. Hirano, T. Zako, S. Katsura, T. Nagamune, A. Mizuno: *Direct Observation of Sliding of Restriction Endonuclease EcoRI on a Single DNA Molecule*, European Biophysics Journal **29**, No. 4–5, 255 (2000)
38. M. Washizu, O. Kurosawa: *Electrostatic Manipulation of DNA in Microfabricated Structures*, IEEE Trans. on Ind. Appl. **26**, 1165–1172 (1990)
39. M. Washizu, T. Yamamoto, O. Kurosawa, S. Suzuki, N. Shimamoto: *Molecular Surgery of DNA Using Enzyme-Immobilized Particles*, Trans. of IEEJ **116** E, 196–202 (1996)
40. K. Hirano, S. Matsuura, Y. Matsuzawa, S. Katsura, A. Mizuno: *Restriction Enzyme Reaction in μm-Sized Area Using Local Temperature Control Technique by Laser Irradiation*, Thermal Science & Engineering **7**, 11–21 (1999)
41. C. Hoyer, S. Monajembashi, K.O. Greulich: *Laser Manipulation and UV Induced Single Molecule Reactions of Individual DNA Molecules*, J. of Biotechnology **52**, 65–73 (1996)
42. X. Meng, W. Cai, D.C. Schwartz: *Inhibition of Restriction Endonuclease Activity by DNA Binding Fluorochromes*, J. Biomol. Struc. & Dynam. **13**(6), 945–951 (1996)
43. Y. Maeda, N. Harada, S. Matsuura, S. Katsura, A. Mizuno: *Local Activation of Restriction Enzyme by Controlling Mg^{2+} Concentration*, Proc. Inst. Electrostatics Japan, Annual Meeting (2000) pp. 107–108

44. K. Hirano, Y. Matsuzawa, H. Yasuda, S. Katsura, A. Mizuno: *Orientation Control, Cutting in Order, and Recovery of Single DNA Molecules Using the Electric Effect Inside Channels on a Glass Substrate*, J. Capillary Electrophoresis and Microchip Technology 006:1/2, 13–17 (1999)
45. D.T. Chiu, A. Hsiao, A. Gagger, R.A. Garza-Lopez, O. Orwar, R.M. Zare: *Injection of Ultrasmall Samples and Single Molecules into Tapered Capillaries*, Anal. Chem. **69**, 1801–1807 (1997)
46. D.T. Chiu et al.: *Chemical Transformations in Individual Ultrasmall Biomimetic Containers*, Science **283**, 1892–1895 (1999)
47. L.F. Braganza, B.H. Blott, T.J. Coe, D. Melville: *Dye Permeability at Phase Transition in Single and Binary Component Phopholipid Bilayers*, Biochim. Biophys. Acta **731**, 137–144 (1983)
48. S. Furukawa, K. Nakatani, H. Misawa, N. Kitamura, H. Masuhara: *Absorption Microspectroscopy of Zinc Tetraphenylporphyrin in an Individual Droplet in Water*, J. Phys. Chem. **98**, 3073–3075 (1994)
49. B. Rotman: *Measurement of Activity of Single Molecules of s-D-Galactosidase*, Proc. Natl. Acad. Sci. USA **47**, 1981–1991 (1961)
50. D.S. Tawfik, A.D. Griffiths: *Manmade Cell-like Compartments for Molecular Evolution*, Nature Biotechnology **16**, 652–656 (1998)
51. K. Jensen: *Smaller, Faster Chemistry*, Nature **393**, 735–736 (1998)
52. S. Katsura, A. Yamaguchi, H. Inami, S. Matsuura, K. Hirano, A. Mizuno: *Indirect Micromanipulation of Single Molecules in Water-in-Oil Emulsion*, Electrophoresis **22**, 289–293 (2001)
53. A.N. Glazer, H.S. Rye: *Stable Dye–DNA Intercalation Complexes as Reagents for High-Sensitivity Fluorescence Detection*, Nature **359**, 859–861 (1992)
54. M. Nakano, J. Komatsu, S. Matsuura, K. Saito, H. Inami, H. Yasuda, K. Takashima, S. Katsura, A. Mizuno: *Development of Amplification of Unknown DNA Fragment and Single Molecule PCR*, Proc. Inst. Electrostatics Japan, Annual Meeting (2001) pp. 141–144

4 Near-Field Optics in Biology

Patrick Degenaar, Eiichi Tamiya

Light microscopy is the single most important tool for cell biology. Visualisation of cells, tissue and microorganisms is taken for granted in most biology labs due to the development of microscopy some 400 years ago. Over the years, lens quality improved until resolutions of around 1 μm were achieved. By 1873 Abbe had developed his fundamental work on diffraction and microscopic imaging. This showed that lateral resolution was limited by the diffraction limit of the wavelength of the light in use. However lens imperfections continued to limit resolution to about half a μm.

The development of the laser in 1960 by Theodore H. Maiman of Hughes Aircraft was the next major step forward [1]. This led to the development of the laser scanning confocal microscope. Compared with an ordinary optical microscope, a good CLSM (confocal laser scanning microscope) improves optical resolution by up to 30%. With a numerical aperture of 1.4 and 550 nm wavelength, a point scanning CLSM can achieve 0.18 μm lateral resolution and 0.59 μm depth resolution. This is, however, at the fundamental limit of what is possible due to the diffractive nature of light.

The diffraction limit, as described by Abbe, imposes a barrier of around $\lambda/2$. Therefore to increase resolution, the wavelengths used would need to go deeper into the UV range. This is problematic due to the high absorption of UV by most materials. Moreover, deep UV is highly ionising and is therefore destructive to cells and cellular components.

De Broglie's work in the 1920s showed that all particles have their own wavelength depending on their momentum. A consequence of this was the development of the electron microscope. The electron microscope family has had a major impact in biology. For the first time it was possible to view biological samples down to the molecular level. Its high scalability and field of view are also a major improvement over the constraints of traditional light microscopy. However, the use of electron beams requires that the electron microscope be used in vacuum conditions. Furthermore, the main elements that make up biological material, namely carbon, oxygen and hydrogen, do not provide good scattering contrast and are electrically insulating. As a result, for high resolution electron microscope imaging, samples must be coated with conductive metal layers. Clearly, samples that are dried, covered in metal, put in a vacuum and then bombarded with high-energy electrons are not in their natural environment. Studies on live cells and other biological structures in aqueous solution are impossible.

The development of the scanning tunnelling microscope (STM) in 1980 brought about a revolution in the form of scanning probe microscopy (SPM) [2,3]. The development of the atomic force microscope (AFM) led to almost atomic level resolution imaging of biological samples. The AFM could for the first time obtain high resolution images of cells [4] and even DNA [3] in both dry and liquid media. However, the nature of the information obtained is its major drawback. The AFM provides information about the force interaction between the tip and the surface. This tends to be only topographic information, and hence the whole array of reporter dyes available to fluorescence microscopy are not applicable.

The SPM revolution did, however, lead to the practical implementation of near-field microscopy. The actual idea of building a near-field microscope to beat the diffraction limit dates back to Synge in 1928. However, at that stage the technology (e.g., computing power and the laser) were not available to realise his concept. The first applications of the electromagnetic near field actually date back earlier to the turn of the century. More specifically there was the problem of the oscillating electrical dipole antenna next to an extended conducting body. This was a major issue in the early days of radio communications.

With the advent of the STM, work was soon underway on a near-field microscope. The first research came from Pohl et al. in 1982 [5]. Since this time near-field microscopy has developed steadily [6,7]. The first functional SNOMs (scanning near-field optical microscopes) came out in the early 1990s [8]. Further development and refinement has continued, and today SNOMs are commercially available from a variety of companies. There are many different implementations but the fundamental interaction is the same. All SNOM systems use a probe, under which the sample is scanned. The near-field interaction between the probe tip and the sample is detected in the far field for each point and an image is built up.

The typical SNOM can achieve resolutions down to a few tens of nanometres within a scan area of 50 μm^2. Sensitivity has been achieved down to the single-molecule level [9–11]. The SNOM is usually integrated into an AFM (atomic force microscope) to provide simultaneous near-field and topographic force imaging. In this book this configuration is called SNOAM [12], but it is also more simply known as SNOM [13] or NSOM [14].

The main disadvantage with SNOM is that it is a serial device, i.e., it takes information from each point or pixel and scans the area, all of which takes time. This contrasts with standard microscopy, which is parallel, i.e., the picture is instantaneous, and variable in real time. Therefore the present trend is to integrate SNOM into optical and confocal microscopy. This will allow the sample to be readily manipulated and viewed at lower resolutions, before being scanned by SNOAM for higher resolution analysis.

4.1 Breaking the Diffraction Barrier

As mentioned in the introduction, the basic idea for overcoming the diffraction limit in microscopy was developed by Synge in 1928. To understand Synge's proposal, two notions that are not normally used in conventional microscopy need to be introduced: these are the notions of far field and near field. The far field is the propagating wave electromagnetic radiation that is readily detectable with our eyes. The optical near field is a non-propagating electric field, which exists at a sub-wavelength distance around a light-scattering or fluorescent particle. If the near field is continuous, then it can be considered as an evanescent field.

All scattered light contains both far-field and near-field components. Normally, however, only the far-field component is detectable due to a very strong attenuation of the near field with distance. The attenuation of the near field is such that there is basically no near-field component in the far field (at a distance greater than or equal to a wavelength). Therefore, in practice, all near-field phenomena must be detected via far-field emissions from near-field interactions. To detect such phenomena, a probe of known scattering properties must be brought close to the sample. The dipole–dipole interaction between the probe and the sample in the near field will result in far-field scattering. The scattered far field depends on the probe–sample interaction and their separation. The resolution will depend on the shape of the near field between the probe and the sample rather than any diffraction limit set by the wavelength of the light used.

Consider Fig. 4.1. A probe with a fluorescent protein at one end is brought close to another type of fluorescent protein on a glass slide. The incident light on the tip will stimulate the first fluorescent protein, which will then emit light with a redshifted spectrum. Consider the case were this redshifted spectrum is set at the excitation peak of the second fluorescent protein. This second protein will then emit a further redshifted spectrum. If the probe is further than a wavelength from the sample, then the interaction between the two proteins results from a far-field wave propagating from the first protein. If the tip is brought to within a wavelength of the sample, then the interaction between the two proteins is a non-propagating dipole–dipole interaction. If the two proteins are brought so close that they bond, then the incident light gets scattered off their combined electronic wavefunction and fluorescence may be quenched. A true bond would probably disrupt the electron transfer processes responsible for fluorescence. Given that the light emitted from an individual protein is extremely weak, then there would only be a detectable signal for the near-field case. Detection would involve the use of filters to filter out all light except the redshifted spectrum from the second protein.

The situation in Fig. 4.1, while good for explanation, is extremely difficult to carry out in practice due to the very low signal-to-noise ratios that would be obtained. A more realistic demonstration can be obtained with nanoscopic light funnel systems. These are known as apertured SNOM [15] systems.

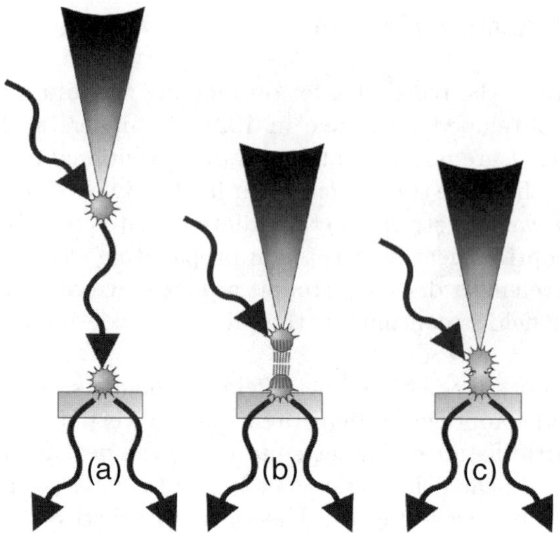

Fig. 4.1. Scattering conditions between a particle on the tip of a probe and a particle on the surface of a cover slide. (**a**) The sample–probe distance is more than a wavelength. Scattering from the probe interacts with the sample in the far field. (**b**) The sample–probe distance is less than a wavelength. The scattering interaction between the tip of the probe and the sample occurs in the near field. (**c**) The two scattering particles are so close that they have bonded. The light scatters off the combined particle

Apertured SNOM probes are single-mode optical fibres that have sharpened tips and a skin-depth metal such as aluminium evaporated on the top. At the very end of the tip, a sub-wavelength aperture (hole) is created. Laser light can thus be inserted into one end of the optical fibre and a near field will exist around the aperture of the other end.

Figure 4.2 compares focal, confocal and apertured SNOM microscopy. In focal microscopy, the whole sample is stimulated by light and a lens is used to focus the bright field, contrast, or fluorescence image into an eyepiece or CCD camera. In the case of confocal microscopy, a laser is focused by a lens onto a point of the sample and the scattering of the laser off that point is detected by a light sensor such as a CCD or a photomultiplier tube (PMT). The laser is scanned over the whole sample to build up an image. The resolution is limited by the aperture, but ultimately by the diffraction limit. In the final case an apertured SNOM probe is brought close to the surface and the near field around the probe aperture interacts with the sample, causing it to emit far-field light. This light is then detected by an avalanche photodiode (APD) or a PMT. The resolution limit in this case is limited to the size of the aperture.

For biological applications, SNOAM acts as an AFM which takes optical data. The optical and topographic images can be compared to better under-

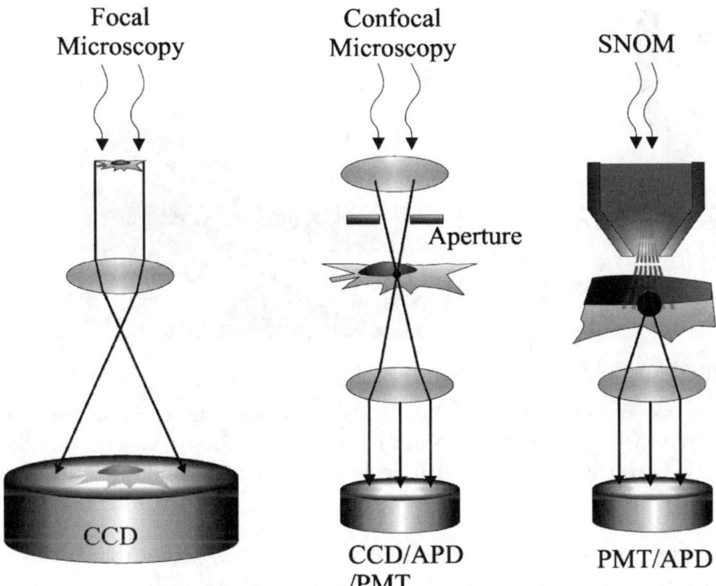

Fig. 4.2. Comparison between the basic optics of different types of microscopy. *Left*: normal parallel optics for focal microscopy. *Centre*: optics for detecting one point of a sample with confocal microscopy. *Right*: optical system for apertured SNOM

stand interesting features. Generally, two types of image are taken: optical transmission images [16,17] and fluorescence images [18–20]. Effects such as polarization [21,22] can also be important tools, but the optical information is generally taken from fluorescence intensity levels.

There are still many theoretical challenges for SNOAM in calculating the optical near-field intensities around probes. Better theory will improve reconstruction and analysis of images and help integrate SNOAM better with standard microscopy. Theoretical analysis must consider the main mechanisms involved in the probe–sample interaction. These include the dipole coupling between the probe and the sample, the polarization of the light, the refractive indexes of the probe and sample, as well as the probe and sample geometries and their separation. Common approaches involve solving Maxwell's equations for the interaction between two dipoles [23–28], though there is still much to be done.

4.2 SNOAM Probe Design

To overcome the diffraction-based resolution limit in visible light microscopy, scanning near-field optical microscopy makes use of a strong localization of light. This localization is commonly achieved using apertured probes, which

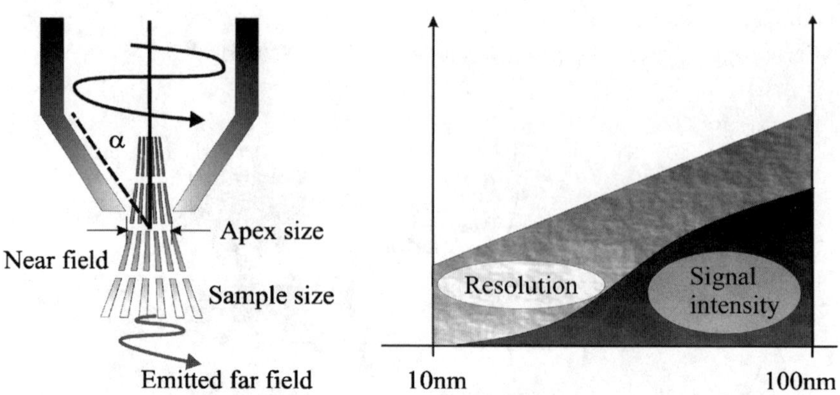

Fig. 4.3. Apex size vs. signal intensity. The optical resolution depends on the size of the aperture, while the topographic resolution depends on the tip size and angle. The signal intensity, however, decreases with decreasing tip angle and aperture size

squeeze light through a narrow metal funnel with an aperture substantially smaller than the light wavelength [29]. This does indeed strongly localize the probing light, ensuring spatial resolutions < 50 nm or $\lambda/10$. However, this gain in resolution is paid for by a severe power loss in the signal. As a result, the limiting factor for the resolution of an apertured probe is the detector's signal-to-noise ratio.

Probe dimensions can be changed so as to increase the throughput for a given aperture. If the probe angle is increased, then the throughput of the probe will also be increased. Most SNOMs are combined with atomic force microscopy for the probe–sample separation. It is therefore desirable to correlate optical data with the simultaneously obtained topography. If the probe angle is increased, its aspect ratio decreases and so too will its topographic resolution. Consider for example an apertured probe with a given throughput and an optical resolution of 100 nm. If it were redesigned to have an optical resolution of 30 nm with a similar throughput, then its ability to penetrate well structures would be very much reduced. Thus a small gain in optical resolution leads to a large loss in topographic resolution. Typical probes have optical resolutions in the range 50–100 nm and topographic resolutions in the range 250–500 nm. AFM probes in comparison, have resolutions of 1–10 nm.

Consider the situation in Fig. 4.4. In this case an apertured probe is entering a well structure. Initially, the probe cannot penetrate due to the width of the base of the probe. However, as the aperture clears the edge of the well, the optical signal increases. Both topographic and optical signals then go flat until the probe starts descent. The optical signal then increases, as the probe gets closer to the glass. In practice, if the well structure is not high, the double dip is not detectable. In that case the profile looks like an inverted version of the topographic signal. It should be noted that neither the optical nor the topographic profile reproduce the original well structure

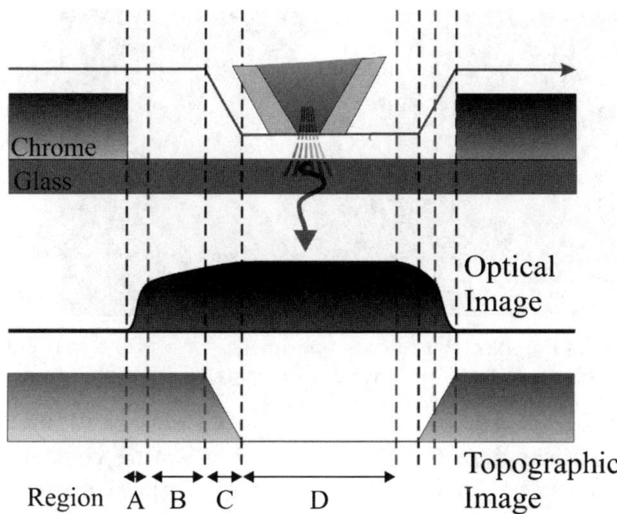

Fig. 4.4. Signal intensity vs. aspect ratio. The aspect ratio is the angle of the tip. High aspect ratios are required for good AFM imaging. Lower aspect ratios are needed for good light throughput. The image above shows the artefacts which occur due to the difference between the size of the aperture and the size of the tip

perfectly. However, using a standard well structure sample, it is possible to calibrate for the tip properties.

There are three forms of apertured probes produced both commercially and in various SNOAM labs. These are straight optical fibre probes, bent optical fibre probes, and micromachined probes. Straight optical fibre probes are more common as they are the easiest to make. Bent optical fibre probes are less common, more time-consuming to make, and hence more expensive to buy. Recently, micromachined probes have become available. These use the same micromachining technology as AFM probes.

Compared to bent probes, straight probes have higher throughput. However, straight probes must be used with shear-force feedback. Shear-force feedback, as will be explained later, inflicts undesirable lateral forces on cellular samples. Moreover, shear-force feedback has lower topographic resolution on rough samples. Bent optical fibre probes have the best feedback properties in water due to their aquadynamic shape [30]. When used in tapping mode [31], they only impart less damaging vertical forces to samples. They do, however, have lower levels of light throughput. Micromachined probes [32–37] are still in the early stages of development. They have the potential to become cheaper and they have higher throughput and thus allow for higher resolutions. On the other hand, they are not as aquadynamic in water due to their flat shape, so pulsed-force feedback, a variant of tapping mode, is necessary.

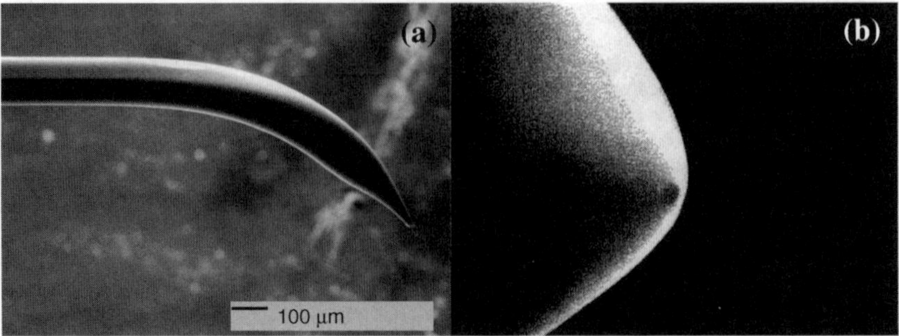

Fig. 4.5. Representative scanning electron microscope images of a probe (**a**) and the tip of the same probe (**b**) made from an optical fibre coated with aluminium

In the case of bent optical fibre SNOAM probes, the spring constant determines the vertical sensitivity and resolution. Typical spring constants are calculated [38,39] to be in the range 2–20 N/m. The Q factor is a quality factor for the resonance curve in the AFM feedback system. It determines the signal-to-noise ratio for the feedback. The resonance frequency must be higher than the background frequencies generated by sound and general noise. The resonant frequencies of other components tend to be around a few hundred hertz, so probe resonance frequencies of the order of tens or hundreds of kilohertz are generally desirable. Q factors are typically 200–600 in the air and 40–200 in water [57]. Resonant frequencies vary from manufacturer to manufacturer.

Bent optical fibres are designed differently, depending on whether they are to be used in air or in water. Initially, both are made from optical fibres, which are chemically etched to make a tip and bent by irradiation from a CO_2 laser. They are then coated with 100–200 nm of metal and an aperture is cut at the tip [40]. Air-operation probes are created from a 125 μm fibre-optic core. The metal covering is aluminium and the probe tends to be thicker and easier to handle. Water-operation probes are created from a 40 mm core. The metal covering is first chrome then an outer layer of gold. These probes are thinner and hence more suitable for use in water [41].

4.3 SNOAM Configurations

Apertured SNOAM systems, i.e., SNOAM systems using apertured probes, have two basic configurations. Light can be passed through the probe and the near field around the probe aperture stimulates the sample. In the second case, the entire sample can be illuminated and the probe used as a sub-wavelength pickup. Finally, in the case of apertured systems, the entire sample is illuminated and a point probe is used as a scattering tool.

In Fig. 4.6, the collection mode SNOAM has a configuration known as photon scanning tunneling microscopy (PSTM) [42] and is generally (though

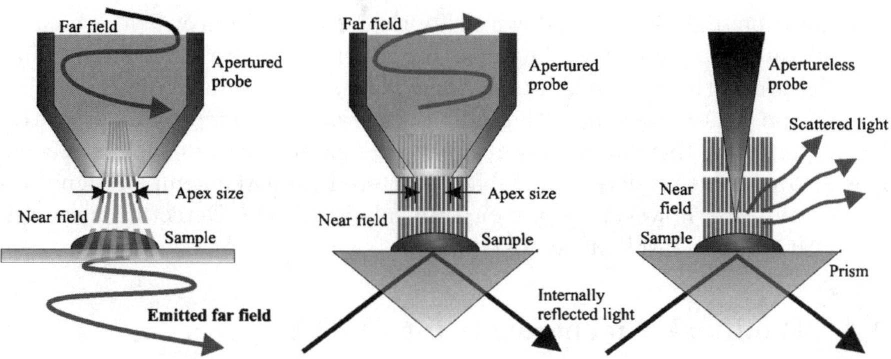

Fig. 4.6. *Left*: light is passed through the probe and the sample is stimulated by the near field around the probe aperture. *Centre*: the sample is set on top of a prism and an evanescent field stimulates the sample. The resultant near field is picked up by the apertured probe. *Right*: the sample is set on top of a prism and an evanescent field stimulates the sample. The resultant near field is scattered by a tip of known scattering properties

not exclusively) used with fluorescence microscopy. In PSTM, the sample is put on top of a prism. A stimulating laser beam is then shone onto the prism at an angle greater than necessary for total internal reflection. While no far field passes through, there is nevertheless an evanescent field. This field will attenuate rapidly and thus will only stimulate fluorescent particles up to around one fifth of a wavelength from the prism surface. The apertured probe then picks up the resulting near field from the fluorophores. Detection of the signal occurs at the end of the fibre with an avalanche photodiode (APD). A filter is inserted before the APD to filter out excitation light. The PSTM configuration cannot be applied to cell biology because stimulation only occurs within a hundred nanometers or so of the surface. Cells have topographies from half to several microns and cannot therefore be studied in this way.

Apertureless systems could possibly adopt the same method as PSTM. In the case of apertureless systems, the tip could act as a vibrating scatterer. The signal would then be picked up as an oscillation ripple on the background signal. Probes with good surface enhancement effects could give useable signal-to-noise ratios. Apertureless SNOAM, like apertured collection SNOAM, suffers from the fact that the whole sample needs to be continuously illuminated over the whole scan. Fluorescent marker dyes used in biology are subject to a phenomenon known as photobleaching, in which the fluorescent signal decays over time. Since SNOAM scans generally take time, continuous photobleaching is clearly a problem.

Illumination mode apertured SNOAM is the most common configuration for cell biology [43–45]. Only one part of the sample is illuminated at a time during scanning and it is possible to image rough samples such as cells. In this configuration, light is passed down the optical fibre and a near field is created around the probe aperture. The near field then stimulates the sample. Far-field light from the near-field interaction will be scattered in all directions. For samples on glass cover slips, the light is generally collected by an objective lens underneath the sample. For samples on opaque surfaces such as silicon, it is also possible to detect light back-scattered from the sample. Signal-to-noise ratios are, however, better for transmission mode. Detection is carried out with either a PMT or an APD.

4.4 Feedback Mechanisms for SNOAM

The optical near field is highly dependent on the probe–sample distance [46]. The tip must be kept within a few tens of nanometres from the sample to get a good signal, without crashing into it. A feedback system is therefore necessary to regulate the probe–sample distance during scanning. There are many different feedback methods such as STM (scanning tunnelling microscopy) [47], lateral shear force [47–49], and contact-mode atomic force microscopy [50]. For biological samples in aqueous environments, tapping mode and shear-force systems are generally favoured.

Straight probes can only be used with shear-force feedback. Bent probes can be used in contact mode and a variety of tapping modes. Contact mode, like shear-force mode applies damaging lateral forces to cellular samples and tapping modes are therefore preferable. There are three types of tapping mode: non-contact tapping mode, intermittent contact tapping mode, and pulsed-force mode. Bent optical fibres work best in both non-contact and intermittent contact modes. Micromachined probes, which suffer from damping effects, work best in pulsed-force mode.

Both tapping and shear-force modes operate under similar principles. The probe is vibrated at its primary resonant frequency and brought to the surface. When the probe nears the surface, the atomic forces between the tip and the sample shift the resonant frequency. The shift in resonant frequency indicates the sample–probe separation and is picked up by the feedback mechanism. Usually, a laser is used to detect the vibration at the tip, but in some cases, especially for shear-force mode, a tuning fork is used. Once the shift in vibrational resonance frequency is detected, it is back-convoluted by the feedback electronics to give the sample–probe distance. The feedback loop then alters the sample height relative to the probe. As in AFM, it is generally easier to move the sample rather than the vibrating probe.

The quality of the resonance curve, also known as the Q curve, determines the probe sensitivity. An ideal Q curve would be very high and very thin so that small shifts in the resonance peak are distinguishible. Thus a Q factor (quality factor) can be attributed to each probe. The Q factor is the height of the resonance curve divided by its width at half the maximum height. The better the Q curve, the easier it is for the feedback electronics to distin-

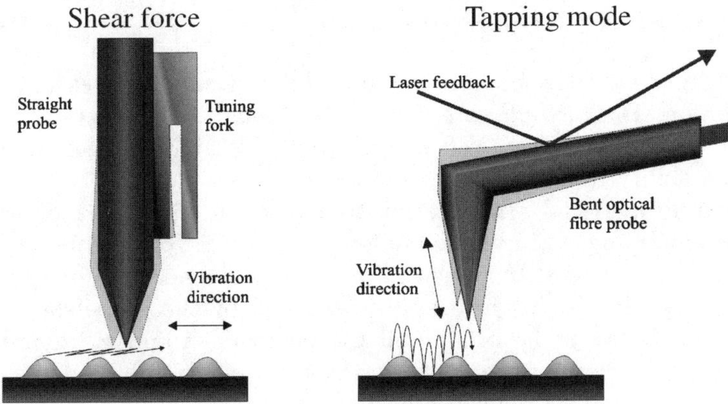

Fig. 4.7. Feedback modes for SNOAM. *Left*: shear-force mode. *Right*: tapping mode

guish between shifts in resonant frequency. Thus the Q factor determines the vertical topographic resolution of tapping probes.

Tapping mode uses bent optical fibre probes, which vibrate vertically relative to the sample. Shear-force mode uses straight optical fibres, which vibrate horizontally relative to the sample. Both have advantages, but in general tapping mode is superior for analysis of biological samples. Shear-force probes have less loss through their fibres, but cause more damage to soft samples. Shear forces cause a lot more damage to cells than gentle tapping. Furthermore, the topographic resolution depends not only on aspect ratio, but also on vibration, making resolution somewhat lower than tapping mode for rough samples. Finally, shear-force systems suffer from more artefacts, especially when used in water. However, there are some situations, such as with IR or UV optics, where bent optical fibre probes cannot be used.

Pulsed-force mode is a hybrid between contact and tapping mode. The probe receives a vibration pulse which lifts the probe before scanning to the next position. Atomic forces between the sample and the probe then cause the probe to bend, and this is detected with a feedback laser. It is not currently very common, but may become more widely used as micromachined probes become more readily available.

Most commercial SNOAM systems have the ability to switch between different types of force feedback. Future applications will use SNOAM not only as an imaging tool, but also as a manipulation tool. To this end, contact mode, or scanning at some of the probe's higher resonant frequencies, can be used to cut or otherwise manipulate the sample under investigation.

4.5 SNOAM in Aqueous Environments

When the first AFM was developed, it required high vacuum and extreme vibrational isolation. Today, while somewhat more difficult, operation in liquid has become routine [51]. SNOAM has inherited this ability to function in liquids and its ability to do so is very important for cell biologists.

SNOAM requires a special configuration to operate in aqueous environments. Aqueous environments involve large forces resulting from the surface tension and viscosity of water. As a result, the entire probe needs to be submerged. In principle, the resolution is still determined by the probe shape, but the interaction between the probe and the liquid environment cannot be ignored. The liquid environment puts other forces on the probe due to viscosity and turbulence. This reduces the signal-to-noise ratio for detecting atomic forces in the feedback system.

In practice the most immediate difference can be seen in the quality of the resonance curve (the Q factor). This reduces considerably, and extra resonance peaks are formed. As a result compared to dedicated micromachined AFM and SNOAM probes, bent optical fibre SNOAM probes have an advantage in their shape. Curved optical fibres are inherently more aquadynamic than flat AFM cantilevers [52–54,60].

A liquid chamber designed for tapping mode in the aqueous environment can be seen in Fig. 4.8. Water and cell culture media are held between the glass plate and an upper window. Both the probe and the sample are immersed in solution.

The main consideration of SNOAM to bio-imaging is not the environment itself, but the samples under inspection. Samples can be classed into three categories: dry, wet static, and alive. For dry analysis, cells are normally 'fixed' [55]. Fixation is a kind of fossilisation process, whereby a polymer

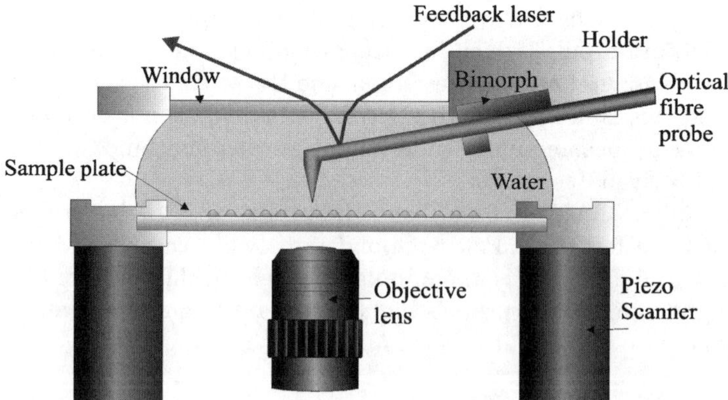

Fig. 4.8. Liquid chamber for SNOAM. Water and specimens are held between the glass plate and an upper window. Both the probe and the sample are immersed in solution

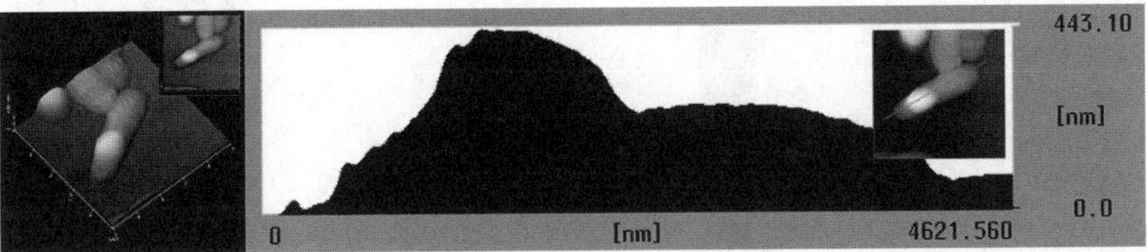

Fig. 4.9. Topographic image and profile of an E. coli cell in a semi-dry state

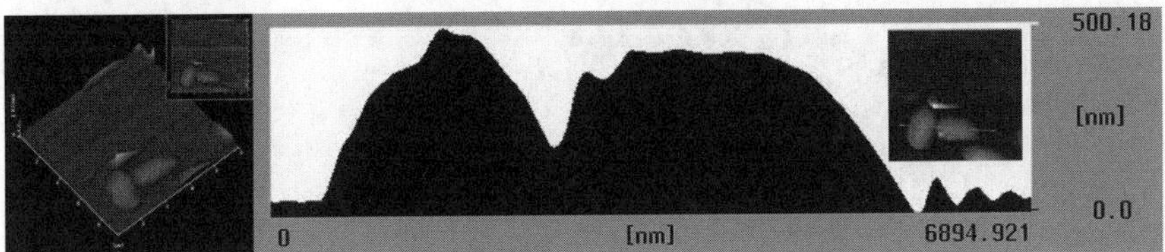

Fig. 4.10. Topographic image and profile of E. coli cells in liquid

such as paraformaldehyde enters the cell and kills it while maintaining the cell structure. This structure is maintained even after drying. It is thus relatively easy to obtain force images from fixed cells when they are dry. *In vitro*, the same fixed samples suffer from the problems associated with the viscosity of water, but generally similar lateral resolutions are possible. In the case of living cells, it is somewhat more of a challenge.

Two problems face analysis of living cells: the problem of keeping the cells alive and well throughout the potentially long period of probe alignment and analysis, and the probe–sample interaction. In fixed cells, the cell membranes do not move. They are like fossilised statues and hence are relatively easy to analyse. Live cells are softer [56], and their membranes generally move. This makes force imaging, very sensitive even to angstrom-scale movements, much more difficult.

Figures 4.9 and 4.10 compare dry and wet E. coli cells. Both cells have been fixed with paraformaldehyde and are thus immobile. The image of the cells in the wet state has less clarity than the one in the dry state. The height of E. coli cells is known to be around 500 nm. In Fig. 4.9, the cell has a height of about 200–300 nm. This is because the drying process deforms the cell shape to make it flatter but wider. In contrast, the wet cells show a height of almost 500 nm.

4.6 SNOAM System Design

SNOAM systems are typically a combination of an AFM system with optics and special mounts for SNOAM probes. The system used in our studies is a Seiko SPA 700 SNOAM [57–61], an all-purpose system which can be used in air or liquid. A photograph can be seen in Fig. 4.11.

Fig. 4.11. The Seiko SPA 700 SNOAM system

The SPA 700 uses bent-type optical fibre probes mounted on a piezoelectric bimorph. The bimorph is used to vibrate the probe, and probe vibration frequency can be scanned over 1–400 kHz, though typical resonance frequencies of Seiko probes are 15–40 kHz.

Vibration detection is obtained through a laser diode and a 4-point photodiode array. The laser bounces off the top of the bent optical fibre and forms a point spread distribution (PSD) on the photodiode array. When neutral, the PSD should be spread evenly between the 4 photodiodes. During scanning, small movements up or down can be detected by a change in the vertical distribution. This can then be calculated and put back through the feedback mechanism. In contact mode, horizontal drag or friction can be measured using horizontal shifts in the PSD. This 'friction force measurement' can be used to carry out surface structure analysis.

As mentioned previously it is much easier to scan the sample relative to the probe than vice versa. The sample stage is moved using a piezoelectric feedback mechanism. This allows scanning up to 70 μm in the horizontal and

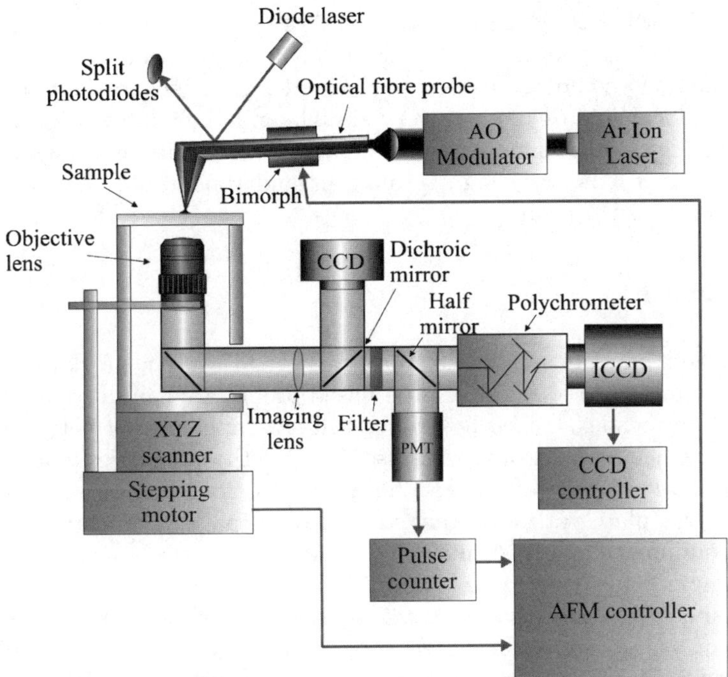

Fig. 4.12. Schematic diagram of the SNOM/AFM system

a few μm in the vertical during a scan. The sample is thus usually placed in position manually, with the aid of the eyepiece before scanning. The tip and mount are then mounted and the stage is scanned roughly up to the tip. As the stage gets closer to the tip, it is scanned more finely and eventually uses the piezoelectric motors for SNOAM scanning. The piezomotors have a z range of 10 μm.

Light input comes from a multiline Ar ion laser, which can be tuned to 457, 488, or 514.5 nm via a polychromatic acoustic optical modulator (AOM). The AOM also synchronously modulates the laser beam to the probe vibration frequency. The laser light is thus pulsed to the point when the probe is nearest the sample. This operation improves the resolution of the optical image [62]. The laser is capable of 150 mW total, but the power at the aperture should be no more than a few mW or it will damage the tip through heating.

The signal can be detected either through transmission or reflection. Either way, it is picked up via a lens and funnelled into either a PMT or an APD. For transparent substrates, the best results are obtained in transmission mode using a 100× oil-immersion-type lens. The signal is then fed into a SNOAM controller unit, which performs fast Fourier transform calculations to determine the signal and send it to the computer for reconstruction.

The designs for the next generation of SNOAM systems adopt a different approach to the one above. The newer designs aim to incorporate the SNOAM into modified confocal microscopes. The design principles between SNOAM and confocal microscopy are very similar and their integration is highly desirable. This will allow imaging from very low resolutions from the microscope right up to the very high resolutions and topographic imaging made possible by SNOAM probes.

4.7 Calibration

The resolution and imaging properties of any SNOAM system are largely dependent on the probes used. Optical-fibre-based probes are handmade and will thus vary in properties. Generally probes will have apertures of between 50 and 100 nm, which will degrade with use. To test the properties of any particular probe, a standard sample can be used. The standard sample we have used is a glass plate with a chequered chromium pattern on top. The evaporated chromium height is 20 nm. The square size varies on different parts of the sample, but a typical size is 0.5 μm^2.

The chromium areas are raised and will therefore give a brighter topographic AFM image. However, chromium is opaque and so all light created from the near field is scattered or absorbed, rather than transmitted. Hence, these areas will be darker in the SNOM image. As a result the two images will be the inverse of each other.

Consider Fig. 4.13. In region A, the probe emerges from the edge of the chrome sample and the optical image increases accordingly. In region B, the probe scans until the edge of the tip can clear the edge. In this case the optical image remains constant. In region C, the probe, having cleared the edge, begins descent. The topographic profile goes down, while the optical profile increases slightly. In region D, the probe hits the bottom of the pit and both the topographic and optical profiles go flat. In practice, the small increase

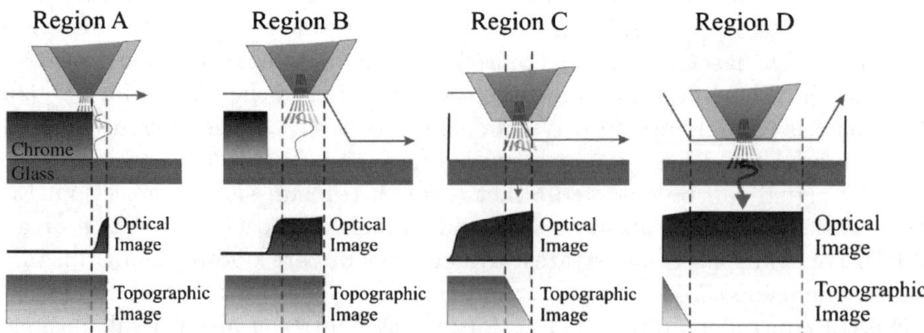

Fig. 4.13. Signal obtained when a SNOAM probe scans over an opaque trench or step structure

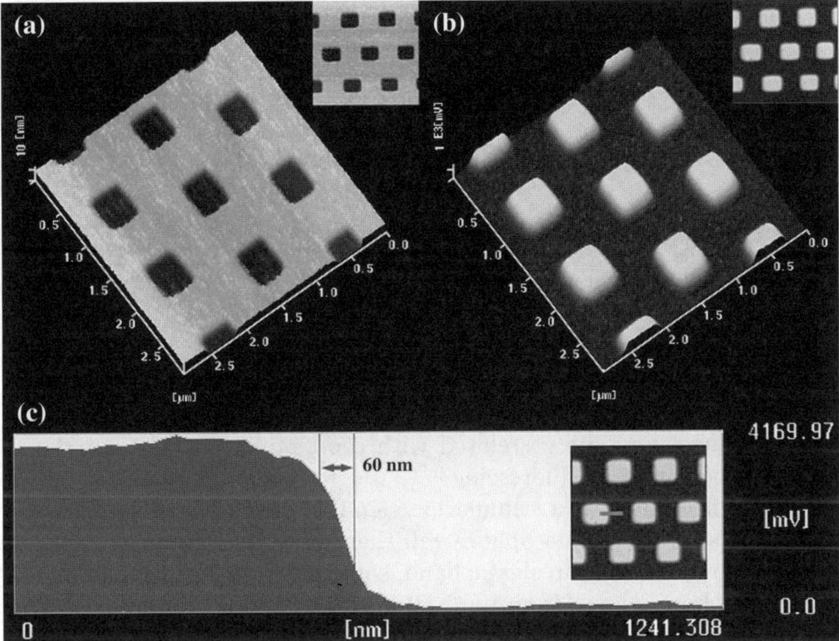

Fig. 4.14. (**a**) Topographic and (**b**) near-field optical images of a standard chromium pattern on a standard quartz plate. (**c**) Optical profile for the bar in (**b**). Scan area 5×5 μm^2

in optical signal between the top and bottom of the pit is not detectable. The small change in signal due to the 20 nm chrome height is not easily seen above the noise level on the main signal.

Figure 4.14 shows the data obtained from a bent optical fibre probe scanning over the chrome sample. In this case an optical resolution of 60 nm was obtained. The topographic resolution cannot be obtained by merely looking at the step width. The topographic image must be compared with the optical image and the distance between the step-up of the optical image and the bottom of the step-down on the topographic image will give the topographic resolution.

4.8 Fluorescence Imaging with SNOAM

Biological samples are normally transparent under the microscope. To overcome this, microscopists use techniques such as phase-contrast imaging. Interesting areas of the sample can be fluorescently tagged, and the fluorescence image correlated with the phase-contrast image. In SNOAM, optical imaging of biological samples generally requires fluorescent tagging. A nonfluorescence image obtained by SNOAM will be a combination of autofluores-

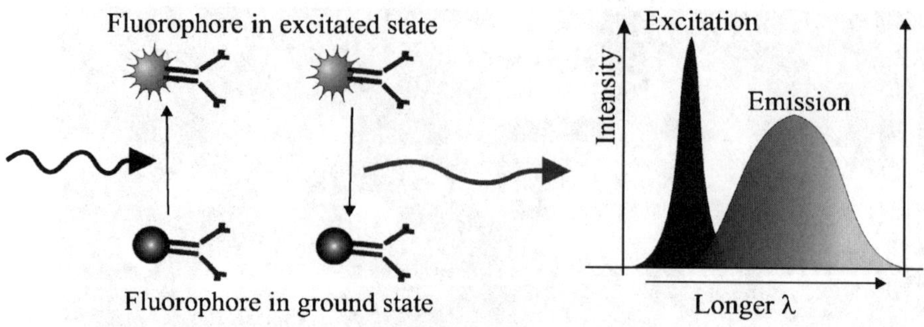

Fig. 4.15. Fluorescence imaging with SNOAM

cence, scattering, and absorption processes. This can be hard to determine. Fluorescence images can be correlated with topographic images to indicate what part of the sample is fluorescing.

In fluorescence imaging, a stimulation signal is used to stimulate the fluorophore. The stimulated fluorophore will then undergo a process analogous to an elastic band. But like an elastic band, some energy is lost in the process. When a fluorophore relaxes from its excited state, a photon is emitted. The emitted photon will be of lower energy than the stimulating photon. A filter can thus be used to separate stimulation photons from emitted fluorescence photons.

The major limitation to all fluorescence imaging techniques is photobleaching. Fluorescence occurs as a result of electron transfer processes in the fluorophore. Stimulation radiation causes the fluorophore to go into an excited state. The fluorophore then relaxes with the emission of a fluorescent photon. Occasionally, stimulating photons put the fluorophore into a doubly or triply excited state and this acts to break the electron transfer bond. This is analogous to a spring being stretched too far. Once a spring goes beyond its elastic limit, it loses its elasticity.

In practice, photobleaching is a process which results in the gradual destruction of fluorophores with time. As a result, the sample under measurement gets fainter and fainter. Different fluorescent substrates have different photobleaching half-lives. But care must be taken with all samples. In apertured SNOAM systems, even though a SNOAM run can take on average 30 min, each pixel area is under illumination for less than a second and the illumination intensity is low. Therefore, samples can usually be scanned many times.

There are a number of methods for fluorescence staining and tagging. The most common methods include antibody tagging, membrane or intracellular staining, and genetic engineering. The method used depends on what needs to be imaged.

In antibody tagging, a primary antibody is directed against the target area of interest. Primary antibodies will have a particular immunoglobin affinity.

Secondary antibodies are attached with fluorophores and then targeted towards the immunoglobin-type of the primary antibody. It is also possible in some cases to use a single antibody, but this requires special preparation.

Membrane and intracellular dyes link directly to their prospective targets. For example Fura-2/AM and Fluo-3/AM can be used to monitor intracellular calcium levels in neuron cells. In this case it is possible to measure intracellular fluctuations by keeping the probe over a single spot [63].

It is also possible to genetically engineer fluorescent proteins such as GFP (green fluorescent protein) into areas of interest on the sample.

4.9 SNOAM Imaging of Fluorescent Beads

To test the fluorescence-measuring properties of a SNOAM system, fluorescent beads can be used. Latex beads of known size and covered with a known fluorophore are useful for fluorescence calibration. The fluorescence signal is very low with the SNOAM system, so an APD (avalanche photodiode) or PMT (photomultiplier tube) must be used in photon-counting mode.

Topographic and fluorescence images of a 100 μm fluorescent bead are shown in Figs. 4.16a and b, where beads are scattered in a PVA [poly(vinyl

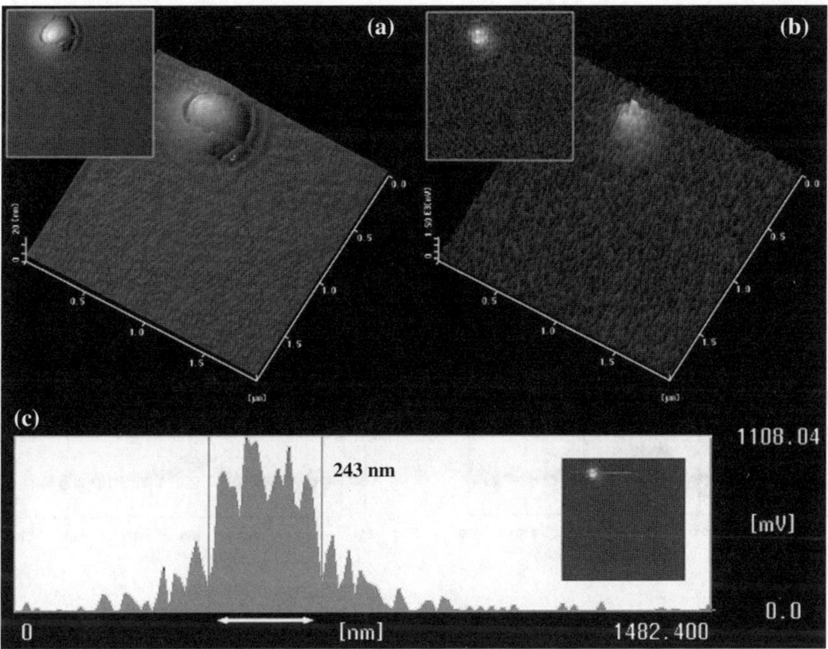

Fig. 4.16. (**a**) Topographic and (**b**) fluorescence images of a 100 nm fluorescent bead on a glass plate spread with a PVA film. (**c**) Fluorescence profile of the bead in (**b**)

alcohol)] film on a glass plate. The topographic image shows the round shape of the bead and wrinkles in the film around it. The fluorescence image was observed with a 488 nm laser beam for excitation and clearly showed the round shape of the fluorescent bead. The fluorescence intensity profile in Fig. 4.16b shows that the half-width of the fluorescence peak of the bead is about 243 nm (Fig. 4.16c). In this case, the diameter of the bead is 100 nm and the aperture of the probe is 100 nm, so that a width of just over 200 nm is reasonable.

4.10 Fluorescence Profiling

SNOAM is also capable of analysing fluorescence spectra. In this case a spectrofluorometer, consisting of a polychromator and an ICCD (intensified charge coupled device) is attached in place of the photon counting head. The sensitivity of the spectrofluorometer is not as high as that of a photon counting head. Thus the aperture has to be bigger and the spatial resolution is consequently lower.

In Fig. 4.17, fluorescence spectra were obtained from 100 nm fluorescent latex beads. The fluorescent dye on the spheres was fluorescein, which has excitation and emission wavelengths of 488 nm and 515 nm, respectively. The

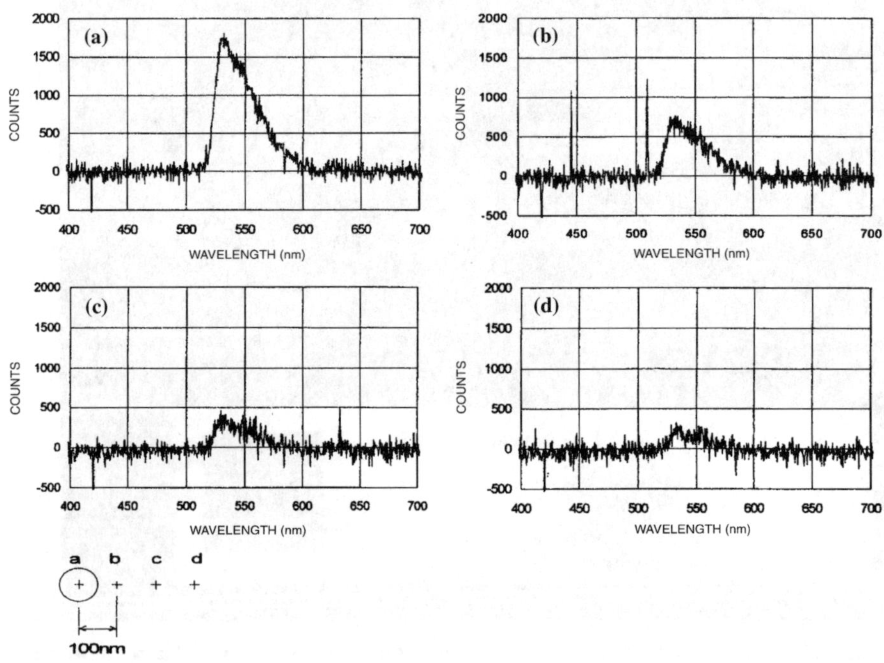

Fig. 4.17. Spectrographs of a fluorescent bead taken in 100 nm steps around the bead with 488 nm excitation

graphs show fluorescence spectrographs when the probe tip was positioned around the fluorescent bead in 100 nm steps. In this experiment, the spectrum window is limited to the range 515–600 nm because the system has a dichroic mirror of 500 nm, a long wave pass filter of 515 nm to cut the excitation light, and a short wave pass filter of 600 nm to cut the laser beam of the optical lever (670 nm). The fluorescence peak weakens with increasing distance from the bead and shows a background signal only at a position 300 nm from the bead centre.

Fluorescence profiling is useful in applications where it is desirable to distinguish between different types of fluorescent dye or autofluorescence.

4.11 SNOAM Imaging of Chromosomes

As genetic vehicles, chromosomes provide the basic material for a large proportion of genetic investigations, from the construction of gene maps and models of chromosome organization to the investigation of gene function. The study of chromosomes has developed in parallel with other aspects of molecular genetics, beginning with the first preparations of chromosomes from animal cells through to the development of banding techniques, which lead to the methods of analytical cytogenetics.

There are 22 pairs of autosomal chromosomes and a pair of sex chromosomes in humans. Many diseases and malformations are a direct result of missing, broken or extra chromosomes. Karyotyping is a simple way to evaluate chromosomes. A karyotype characterises the metaphase chromosomes of an individual cell, arranged in pairs and sorted according to size. Cytogenetics is a method of karyotyping which can recognize and identify gross chromosomal abnormalities.

Characterization of the karyotype can be defined in terms of number, length, centromere position, and availability of satellites. The centromere is the point or region on a chromosome to which the spindle attaches during mitosis and meiosis. When chromosomes are stained by methods that do not produce bands, they can be arranged into seven readily distinguishable groups based on descending order of size and position of the centromere.

- Group 1–3 (A). Large metacentric chromosomes readily distinguished from each other by size and centromere position.
- Group 4–5 (B). Large submetacentric chromosomes which are difficult to distinguish from each other.
- Group 6–12, X (C). Medium-sized metacentric chromosomes. The X chromosome resembles the longer chromosomes in this group. This large group is the one which presents the main difficulty in identification of individual chromosomes without banding techniques.
- Group 13–15 (D). Medium-sized acrocentric chromosomes with satellites.
- Group 16–18 (E). Relatively short metacentric chromosome (#16) or submetacentric chromosomes (#17 and 18).

- Group 19–20 (F). Short metacentric chromosomes.
- Group 21–22, Y (G). Short acrocentric chromosomes with satellites. The Y chromosome is similar to these chromosomes, but bears no satellites. The Y chromosome is similar to these chromosomes, but bears no satellites.

Basic karyotype analysis is based on the results of several conferences (International System for Human Cytogenetic Nomenclature, ISCN) established in Denver, USA, in 1960. Chromosome number and morphology were decided by banding or non-banding techniques. When chromosomes are stained by methods that do not produce bands, they can be arranged into the seven readily distinguishable groups (A–G) based on descending order of size and position of the centromere. Usually, the centromere of a metacentric chromosome appears as a pronounced minimum in either the width or shape profile, whereas that of an acrocentric chromosome is represented only by a smaller-than-usual gradient at one end of the profile.

Chromosomes are only a few microns in length so that high resolution microscopy is desirable for karyotyping. Laser scanning confocal microscopy (LSCM) can achieve around 200 nm as mentioned earlier. A dedicated AFM can obtain nanometric resolutions, but can only obtain topographic information. Using fluorescence SNOAM imaging, it is possible to verify topographic images and correct for any artefacts.

In this research, human metaphase chromosomes were imaged and identified using SNOAM [64,65]. The metaphase chromosomes were derived from human B cell lymphoblastoid line RPMI1788. The cells were grown in RPMI 1640 medium with 10% fetal calf serum (FCS) at 37°C. Metaphase chromosomes were obtained after addition of colcemid (with a final concentration of 0.05 mg/ml), which synchronized the cell cycle. Synchronized cells were harvested by centrifugation (500*g*, 5 min). Chromosomes were then prepared by the 'surface-spreading whole-mount technique', i.e., after centrifugation, collected cells were placed on the clean surface of distilled water with a clean platinum loop. Here, the cells burst due to osmotic pressure and the cell debris spread out rapidly over the water surface. The chromosomes were transferred to a glass cover slip by contact with the surface of the water. The sample was then dried.

SYBR Green I (excitation peak 497 nm, emission peak 520 nm), which binds to DNA strands, was used as a fluorescent intercalater. SYBR Green I is the most sensitive stain available for detection of nucleic acids. Less than 20 pg of double-stranded DNA can be detected in a single band of a SYBR-Green-I-stained gel using 254 nm epi-illumination.

The samples were then imaged using the Seiko SPA700 SNOAM system, described earlier. Bent optical fibre probes were operated in air using cyclic contact mode. The fluorescence images were detected at 520 nm emission with 488 nm argon laser beam excitation.

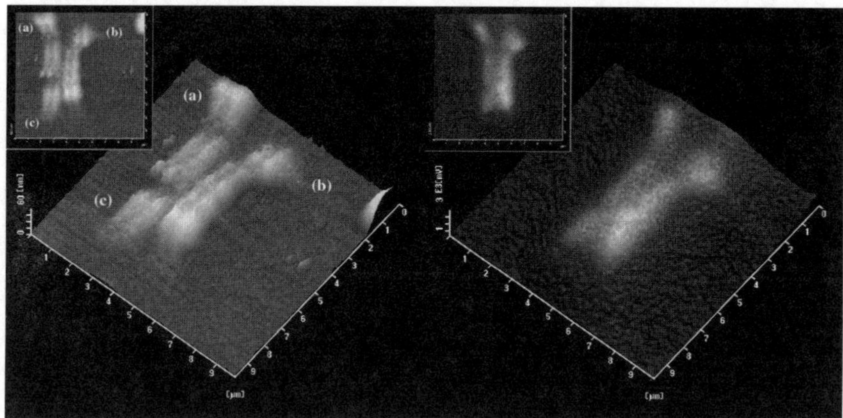

Fig. 4.18. Human metaphase chromosomes imaged under SNOAM. *Left*: dynamic-mode AFM (DFM) image displaying the topography of human metaphase chromosomes on a glass cover slip in air. *Right*: near-field fluorescence image of the chromosome corresponding to the topographic image

Figure 4.18 shows a typical topographic and fluorescence image of human metaphase chromosomes. Topographic images clearly indicate the duplicate structure of the metaphase chromosome, while the fluorescence images are a somewhat different shape. This probably results from the combination of SYBR Green I and chromosome DNA. SYBR Green I is thought to bind non-covalently to the phosphate backbone of the DNA molecule.

The chromosome can be karyotyped as follows. Chromosome (b) was measured as approximately 7 µm. Its centromere is on its left side. Chromosome (b) can therefore be characterized as either #4 or #5 in group C. Similarly, chromosome (a) can be characterized as part of group E or F, and numbered 16, 19, or 20 because it is about 6 µm in length and has a metacentric centromere. Chromosome (c) is approximately 4 µm in length and has an acrometric centromere. It can therefore be characterized as a Y chromosome.

4.12 SNOAM Imaging of Recombinant Bacterial Cells Containing a Green Fluorescent Protein Gene

Green fluorescent protein (GFP) was originally isolated from the jellyfish Aequorea victoria and has become a useful reporter molecule for monitoring gene expression and protein localization in vitro and in real time [66,67]. Normally, proteins need cofactors for fluorescence like NADH or flavine mononucleotide [68,69]. The novelty of GFP is that its fluorescence does not require these [66]. Moreover, since the genetic sequence of GFP is known, it can be engineered into interesting areas of the cell. Native GFP has an excitation peak at 395 nm, but has a broad excitation spectrum, which extends up to 500 nm.

The E. coli (Escherichia coli) cell is the most fundamental cell in molecular biology. Its properties are well understood and it is a good target for genetic modification. However, E. coli cells are approximately 2 μm long and 1 μm wide. Their imaging therefore requires high resolution.

In this research [70,71], spatial analysis was carried out on recombinant E. coli containing the GFP gene. The aim was to look at how GFP molecules give fluorescence activity inside E. coli cells. GFP used in this study was based on the pGFP [66] gene containing the sequence of the wild-type GFP. The E. coli strain JM109 was transformed with pGFP [72] and grown at 37°C. Isopropyl-β-D-thiogalactopyranoside (IPTG) was used as an inducer for effective expression of GFP in the E. coli, because IPTG activates the lac z promoter upstream of the GFP. The transformed bacteria was plated on media containing IPTG and then incubated at 37°C. The following day, the recombinant E. coli were counted by conventional epi-fluorescence microscopy.

For observation with SNOAM, E. coli cells were immobilized on a glass cover slip and dried. The E. coli cells were then chemically bound to the glass cover slip for observation in aqueous solution. Chemical bonding was carried out by pre-treating the glass with 5% γ-aminopropyltriethoxysilane/acetone solution and then 5% glutaraldehyde solution. The E. coli suspension was left on the treated glass surface for 30 min and then replaced by water. The SNOAM system was used in cyclic contact tapping feedback mode. 458 nm and 488 nm lines from the Ar ion laser were used for stimulation. A photon-counting module was used for fluorescence detection, while a fluorospectrometer was used to obtain a spectrum.

Figure 4.19 shows images taken from non-fluorescent and mixed fluorescent E. coli cells. The topographic image clearly shows the outline of four in-

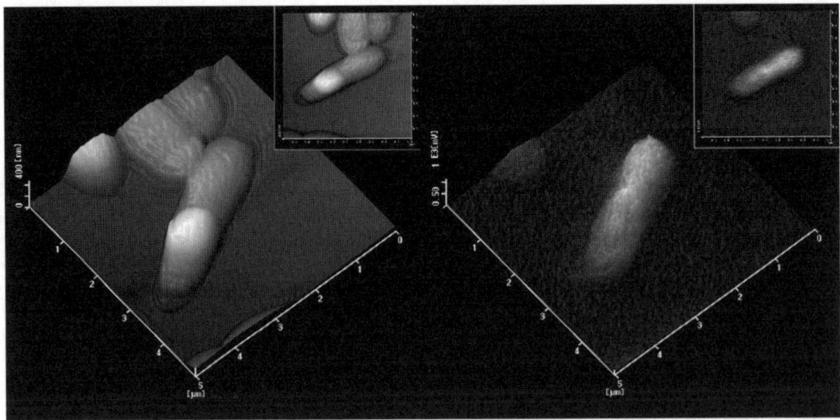

Fig. 4.19. SNOAM images of a mixture of fluorescent and non-fluorescent yeast cells. *Left*: topographic image. *Right*: near-field fluorescence image

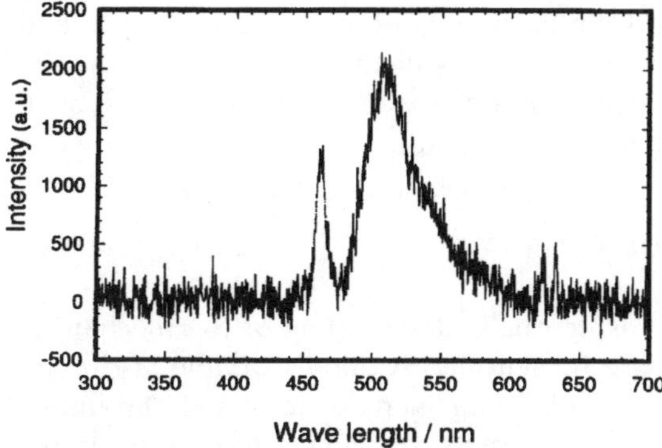

Fig. 4.20. Fluorescence spectrum recorded by positioning the optical fibre probe above the bright area of the E. coli cell shown in Fig. 4.19. The excitation wavelength was 458 nm

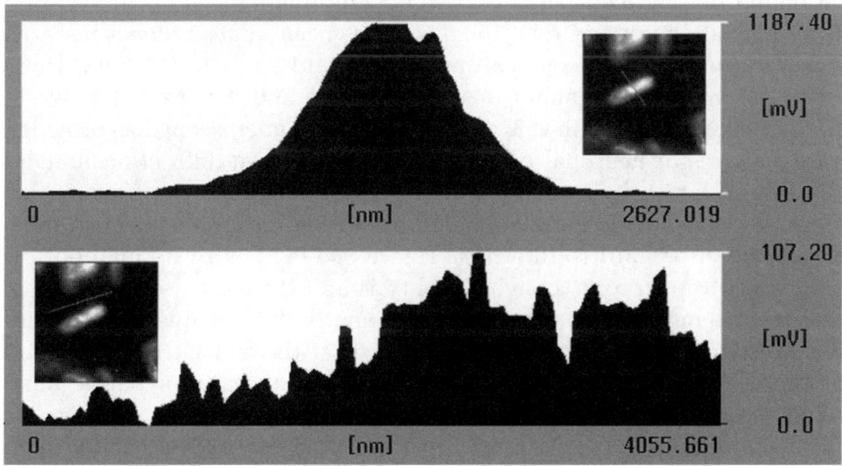

Fig. 4.21. Comparison between fluorescence intensity profiles of bright (*upper*) and dark (*lower*) E. coli cells. Each profile was measured for the bar in the fluorescence image

dividual cells. However, the fluorescence image only shows two. This clearly shows the power of a SNOAM system for distinguishing objects compared with a more conventional AFM.

A good check to make sure that the fluorescence is indeed from GFP is to examine the fluorescence spectrum. Figure 4.20 shows a fluorescence spectrum obtained from a bright region in the cell. The probe aperture in this case was around 100 nm.

Comparing topographic and fluorescence images, it is clear that individual E. coli expressed different fluorescence intensities. Figure 4.21 shows intensity profiles of bright and dark cells. The upper cell expressed 10 times higher fluorescence than the lower cell. Cellular fluorescence expression depends on regular transcription, translation, and post-modification of proteins. The time required for post-translational maturation of the GFP chromophore can be lengthy and appears to vary from cell to cell [73].

4.13 Imaging of Neurons

There is a great interest in increasing our understanding of the mechanisms involved in information processing in neurons. A good example is the area of LTP (long term potentiation). LTP is an increase in signal throughput efficiency which the synapse undergoes after receiving a large repitition of signals. This increase in throughput efficiency is largely believed to be the basis for memory in the brain [74,75]. However, the exact mechanisms of LTP remain unclear. It is believed that structural changes such as the concentration of n-methyl-D-aspartate (NMDA) glutamate receptors at synaptic interfaces play an important role [76]. There has been some excellent research using electron microscopy, confocal microscopy and AFM to this end. However, electron microscopy cannot observe live cells and it is thus difficult to analyse live cell function with this method. Confocal microscopy has provided beautiful pictures of neuronal cells but suffers from the diffraction limit of light, on top of which it is not possible to obtain simultaneous topographic images. AFM can provide extremely high resolution images, but topographic images alone can be hard to interpret. Hence SNOAM, with its high optical resolutions and topographic imaging ability is an extremely useful tool.

Neuron imaging is different to chromosome or E. coli imaging because neurons are larger. The overall size of a neuron with its dendrites and axon can be tens of microns. However, interesting features such as synaptic junctions and receptors are on the nanometer scale. It is therefore useful to be able to do initial scans over a large area and then successively zoom in to more interesting areas. The aspect ratio and size of the probe tip play a large part in the final topographic image. Therefore optical transmission images can be used to check topographic images for artefacts. Figure 4.22 shows a picture of two intertwined neurons. The initial scan is 72 μm followed by a zoom on the overlapping area of the neurons. The second scan zooms in on an area just below the neuron overlap. In this area the neurites are up to 300 nm in height. To check for artefacts in the topographic image, an optical transmission image is very useful. This image depends on the near-field interaction with the surface and the scattering profile. Generally, steep edges of the cell appear dark whilst flatter areas appear bright (even brighter than for the flat glass).

Neurons where extracted from the hippocampal region of chick embryos [77]. Embryos were extracted from chicken eggs grown typically for 8–12 days in an incubator. Cells where incubated in DMEM medium with 10% BFS at 37°C and 5% CO_2 for 3–7 days to form neural networks. The cells in the

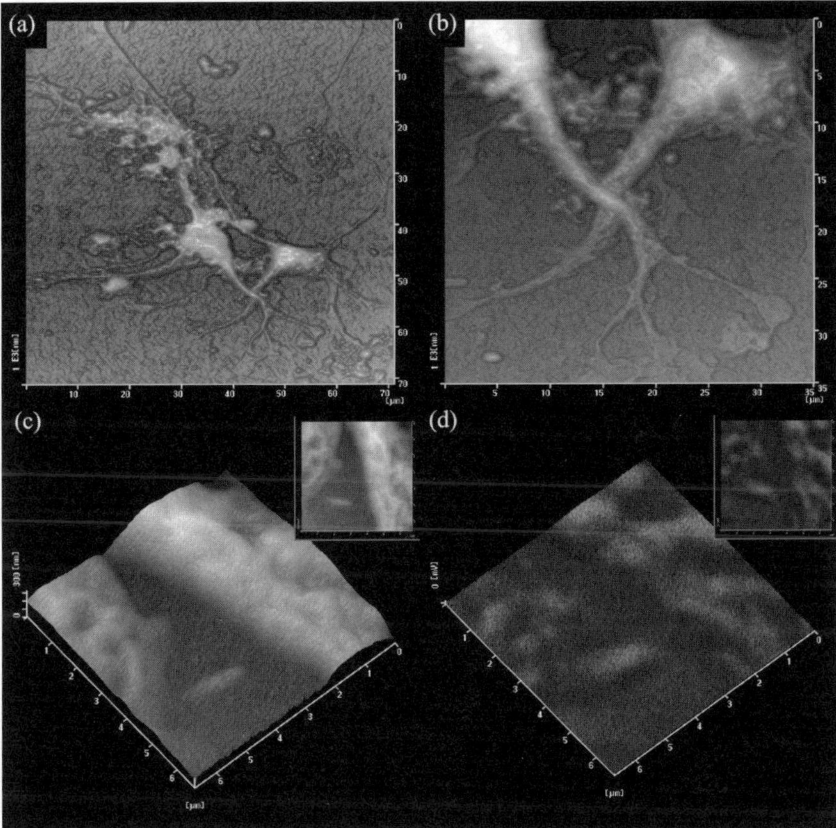

Fig. 4.22. Image of two intertwined neurons. (**a**) Initial topographic scan. (**b**) The second scan zooms in on an area where the dendrites overlap. Note the growth cones, clearly visible on the *bottom right*. (**c**) Topographic image of the membrane just below the overlap. (**d**) Optical image of (**c**). Note that the darker regions indicate either no cell or a steep angle with respect to the cell

above experiment were then fixed with 10% paraformaldehyde and dried. The SNOAM system for the above experiment was used in tapping mode with non-fluorescent optics.

Given the resolution of the SNOAM system, it would be interesting to monitor not just physical features, but transient neurotransmitter release. A SNOAM scan typically requires 20 min or more and is obviously too slow to record transient signals over the whole field. However, if the probe is maintained over an interesting spot, it is possible to monitor transient signals at that position. Candidates for this research would be calcium imaging in the membrane with Fluo-3AM [63] or glutamate imaging with Amplex Red [78]. Glutamate is a neurotransmitter linked with LTP formation and hence

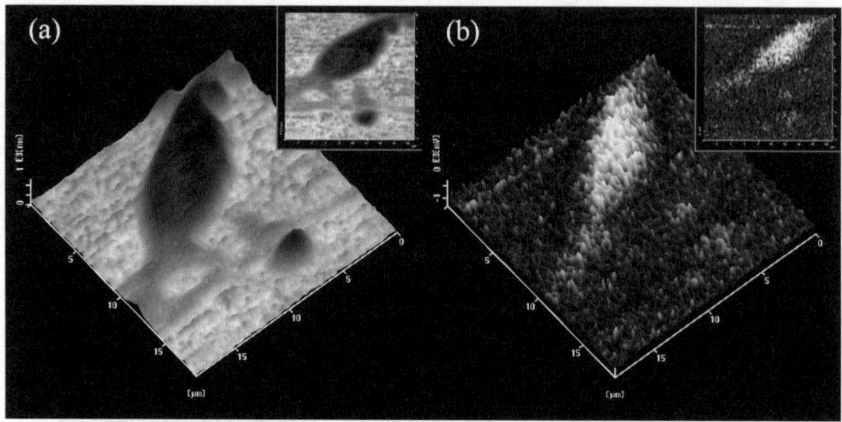

Fig. 4.23. SNOAM topographic image of a neuron soma and neurites (**a**). (**b**) Optical fluorescence image of glutamate release

memory in the brain. Imaging its release with SNOAM could therefore be very useful.

For analysis, the culture medium was replaced with a reaction buffer consisting of 170 mM NaCl, 3.5 mM KCl, 0.4 mM KH_2P0_4, 20 mM HEPS, 5 mM $NaHCO_3$, 5 mM glucose, 1.2 mM Na_2SO_4, 1.2 mM $MgCl_2$ and 1.2 mM $CaCl_2$ in 1 L of MilliQ water. Then 50 ml of 1 unit/ml peroxidase, 50 ml 0.25 unit/ml glutamate oxidase and 10 ml 50 unit/ml Amplex Red were added to the neuron culture. KCl was then added to the neurons to stimulate action potentials and hence glutamate release. The release of glutamate has a chain affect which causes the 10-acetyl-3,7-dihydroxyphenoxazine form of Amplex Red to turn into the fluorescent resorfine form. After thirty minutes, 10% formalin solution was added to fix the neurons. Over time, resorfine will diffuse through the solution and raise the general fluorescence intensity. However, enough remains around the neurons to show contrast. For these results, SNOAM was operated in cyclic contact mode, with 545–580 nm emission filters to obtain the fluorescent signal.

The technique is limited by the use of Amplex Red [78] rather than the SNOAM system. Calcium dyes such as Fluo-3/AM and Fluor-2/AM can be used to monitor calcium transients in real time, but at present there is no equivalent for glutamate.

Apart from its use as an imaging system, SNOAM has other advantages. Added functionality can be achieved by the possibility of stimulating neurons, either by pressure or electrically with the SNOAM probe, or by using the probe as a light source to release caged molecules locally for stimulation. The SNOAM technique will thus continue to develop as a useful tool in neurobiology.

4.14 Future Development of SNOAM

The SNOAM technique still requires a lot of development. Biologists desire high resolution imaging of cellular structures, but also instruments that are easy to use and have high functionality. Initial SNOAM systems were separate systems, not integrated into commercial microscopes. This meant that good far-field imaging had to be done separately from SNOAM. Given the small increase in lateral resolution with these systems, their adoption by biology labs has been slow.

Good confocal microscopes can achieve lateral resolutions of 200 nm. Present SNOAM systems typically achieve 100 nm lateral resolution but have lower signal-to-noise ratios. Confocal microscopy is very easy to use and can capture images in milliseconds. In contrast, SNOAM is a difficult instrument to use and it can take 30 min to get a comparable image. SNOAM has many advantages over normal microscopy, however. The vertical sensitivity is a few tens of nanometres compared to half a micron for conventional microscopy. There are also certain tasks such as manipulation that are not possible with confocal microscopy.

In general, SNOAM needs to integrate itself better with conventional microscopy. Biologists want to take normal far-field images and only use SNOAM once they have found an interesting area on their sample. Better optical designs will allow for phase, Hoffman and interference contrast imaging before scanning. Fluorescence imaging can be carried out in the far field before scanning to find the optimum areas of the sample. Laser scanning and sample scanning laser confocal microscopy will also speed up the image capture process before SNOAM imaging. Many more improvements, especially in probe design [79–81] and integration with laser systems [82,83], coupled with easier use will lead to more laboratories adopting SNOAM for high resolution analysis.

In the future, SNOAM systems will not only integrate themselves better with confocal microscopy, but they will also add functionality. SNOAM is part of the SPM family. Therefore an integrated microscope system should also be capable of using other SPM probes such as high resolution AFM probes. Other experiments such as two-photon fluorescence [84,85] can be done if the signal is good enough. Sample manipulation will provide SNOAM with a 'killer application' that is not possible with conventional microscopy.

The optical resolution of the SNOAM tip is limited by aperture size. However, the smaller the aperture, the weaker the signal. This constitutes one practical limit. The other is in producing probes with small apertures. Present optical fibre probes can be produced with apertures down to 50 nm, but getting lower is difficult. Micromachined probes offer the promise of higher signal throughput, smaller apertures, and cheaper, batch fabricated probes. Micromachined probes also offer possibilities for built-in piezo feedback detectors [86], making SNOAM easier to use. Present developments in quantum well detectors could provide great advances in single-photon detection over photomultiplier tubes and avalanche photodiodes. If this is possible, then the resolution of apertured probes could be brought down to around 10 nm. At this level, there is a 20-fold lateral resolution increase over confocal microscopy which will start to be useful for observing DNA [87,88].

4.14.1 Apertureless SNOAM

In the physics community, a large amount of research is being carried out into apertureless SNOAM. Apertureless SNOAM uses a very sharp metal spike rather than a bent optical fibre probe. Radiation is provided either by an evanescent field, as in collection-mode SNOAM, or shone directly onto the sample. If the tip is modulated, then the scattering resulting from the near-field interaction between the tip and sample is also modulated. This modulated signal can then be distinguished from the constant background signal. In a non-metal probe, this modulated signal would be very low compared to the background signal. However, surface plasmons at the tip can cause enough enhancement to provide a detectable signal [89]. Another method of contrast is to employ fluorescence resonant energy transfer [90] techniques to use emitted fluorescence from fluorophores on the tip to stimulate fluorophores on the sample. Although these techniques are still in their infancy, when developed they will bring SNOAM down to atomic resolutions like its AFM and STM siblings.

However, apertureless SNOAM will not be as directly useable in the biological community as in the physics community. The sample must be continuously illuminated during the scan, leading to photobleaching of the fluorophores used. Sample preparation may also be more complicated, as gold colloids may need to be bound near fluorescent sites in order to achieve enough signal through surface enhancement.

4.14.2 Vibrational Spectroscopy

At present most contrast in biological SNOAM is carried out using fluorescence imaging. Interesting areas can be tagged and thereby imaged. Another method available to biologists is vibrational spectroscopy. This divides into two parts: Raman shift spectroscopy and IR spectroscopy. Basically, these both look at vibrations in the molecules under analysis. Raman shift peaks and IR absorption peaks will indicate the presence of certain types of bond and hence indicate the molecules under analysis. In theory, if combined with SNOAM, these techniques could form a chemical map on the nanometric scale.

Raman spectroscopy is notoriously ‘light hungry’. The signal is usually quite low. When coupled to the usual throughput efficiencies of 10^{-5} for apertured SNOAM, this makes it still more difficult. Nevertheless, this area is now creating a lot of interest, and more and more papers are coming out in this area. The ultimate goal of this research is to develop an apertureless Raman–SNOAM system which can detect down to the single-molecule level.

IR spectroscopy is not as 'light hungry' as its Raman counterpart. However, it uses light in the IR region. As a result, materials such as glass and water which have high absorption coefficients complicate the optics somewhat. Clever optical and system design can nevertheless get round these challenges [91,92]. While a typical resolution of 500 nm ($\lambda/20$ for $\lambda = 10$ mm) does not initially seem impressive, this is nevertheless a major boost to chemical mapping. Most research round the world in this area involves large and expensive free electron lasers (FEL) [93] based at synchrotrons. However, FELs are getting smaller and will come within the financial reach of the larger laboratories over the next decade. However, laser scanning allows for much faster image capture than the sample scanning method with SNOAM.

4.14.3 Competition for SNOAM

The SNOAM technique provides the only physical method that can break the diffraction barrier. However, there are other possibilities. A technique using destructive interference called STED (stimulated emission depletion) [94] could bring the resolution of fluorescence laser scanning confocal microscopy down to 50 nm. The technique operates by quenching the emission laser spot diffraction pattern with an out-of-phase stimulated depletion spot diffraction pattern. The subsequent stimulation spot is substantially reduced and the resolution is thereby increased. This technique only works with fluorescence microscopy and requires a generally expensive tuneable laser.

4.15 Conclusion

In the early stages of AFM and STM, these microscopes were used simply as analysis tools. Later they also came to be used as manipulation tools. SNOAM will also take this path. Because of its dual AFM and near-field capabilities, SNOAM will prove to be an extremely useful tool for manipulation of the nano-environment in biological samples. The AFM side can be used for cutting and putting pressure on small points. The SNOAM side can be used for visual stimulation, heating, and a whole host of effects such as release of caged molecules. These capabilities will allow SNOAM to be a tool for much interesting research in years ahead.

References

1. T.H. Maiman: *Stimulated Optical Radiation in Ruby*, Nature **187**, No. 4736, 493–4 (6 August 1960)
2. L.A. Bottomley, P.N. First, J.E. Coury: *Scanning Probe Microscopy*, Anal. Chem. **68**, 185R–230R (1996)

3. R.T. Dame, N. Goosen, C. Wyman: *H.NS Mediated Compaction of DNA Visualised by Atomic Force Microscopy*, Nucleic Acids Res. **28** (18), 3504–3510 (2000)
4. R. Eschrich, T.A. Kell, G.L. Kumar: *Atomic Force Microscopy on the Olfactory Dendrites of the Silkmoths Antheraea Polyphemus and A. Pernyi*, Cell Tissue Res. **294**, 179–185 (1998)
5. D.W. Pohl: European patent application number 0112401, filed 27 December; US patent 4,604,520, filed 20 December 1983
6. E. Betzig, T.D. Harris, J.K. Trautman, J.S. Weiner, R.L. Kostelak: *Breaking the Diffraction Barrier: Optical Microscopy on a Nanometric Scale*, Science **251**, 1468–1470 (22 March 1991)
7. A. Lewis: *The Optical Near-Field and Cell Biology*, Cell Biology **2**, 187–192 (1991)
8. E. Monson, S. Smith, G. Merrit, J.P. Langmore, R. Kopelman: *Implementation of an NSOM System for Fluorescence Microscopy*, Ultramicroscopy **57**, 257–262 (1995)
9. E. Betzig, J.K. Trautman: *Near-Field Optics: Microscopy, Spectroscopy, and Surface Modification Beyond the Diffraction Limit*, Science **257**, 189–195 (10 July 1992)
10. M.F. Garcia-Parajo, G.M.J. Segers-Nolten, J.A. Veerman, B.G. de Grooth, J. Greve, N.F. van Hulst: *Visualising Individual Green Fluorescent Proteins with a Near-Field Optical Microscope*, Cytometry **36**, 239–246 (1999)
11. R.C. Dunn, S.A. Joyce, E.V. Allen, G.A. Anderson, X. Sunney Xie: *Near-Field Fluorescence Imaging of Single Proteins*, Ultramicroscopy **57**, 113–117 (1995)
12. E. Tamiya, A. Hashigasako, S. Iwabuchi, Y. Murakami, T. Sakaguchi, Y. Morita, K. Yokoyama: *Scanning Near-Field Optical/Atomic Force Microscope (SNOAM) for Biomedical Applications*, Proc. SPIE BIOS 1999 **3607**, 42–49 (January 1999)
13. S. Kirstein: *Scanning Near-Field Optical Microscopy*, Current Opinion in Colloid & Interface Science **4**, 256–264 (1999)
14. H. Shiku, R.C. Dunn: *Near-Field Scanning Optical Microscopy*, Anal. Chem. **71**(1), 23A–29A (1999)
15. C. Obermuller, G. Kolba, G. Abstreiter, K. Karraia: *Transmitted Radiation Through a Subwavelength-Sized Tapered Optical Fiber Tip*, Ultramicroscopy **61** (1–4), 171–177 (1995)
16. G.A. Valaskovic, G.H. Morrison, M. Holton: *Image Contrast of Dielectric Specimens in Transmission Mode Near-Field Scanning Optical Microscopy: Imaging Properties and Tip Artefacts*, J. of Microscopy **179** (1), 29–54 (1995)
17. P.J. Valle, J.-J. Greffet, R. Carminati: *Contrast Mechanisms in Illumination-Mode SNOM*, Ultramicroscopy **71**, 39–48 (1998)
18. H. Muramatsu, K. Nakajima, N. Chiba, T. Ataka, M. Fujihira, J. Hitomi, T. Ushiki: *Fluorescence Imaging and Spectroscopy of Biomaterials in Air and Liquid by Scanning Near-Field Optical/Atomic Force Microscopy*, Scanning Microscopy **10** (4), 975–982 (1996)
19. T.H. Enderle, D.S. Chemla, T. Ha, S. Weiss: *Near-Field Fluorescence Microscopy of Cells*, Ultramicroscopy **71**, 303–309 (1998)
20. P. Nagy, A.K. Kirsch, A. Jenei, J. Szollosi, S. Damjanovich, T.M. Jovin: *Activation-Dependent Clustering of the ErbB2 Receptor Tyrosine Kinase Detected by Scanning Near-Field Optical Microscopy*, J. of Cell Science **112**, 1733–1741 (1999)

21. K. Nakajima, N. Chiba, Y. Mitsuoka, H. Muramatsu, T. Ataka, K. Sato, M. Fujihira: *Polarization Effect in Scanning Near-Field Optical Atomic Force Microscopy (SNOM/AFM)*, Ultramicroscopy **71**, 257–262 (1998)
22. S. Werner, C. Mihalcea, O. Rudow, E. Oesterschulze: *Cantilever Probes with Aperture Tips for Polarization-Sensitive Scanning Near-Field Optical Microscopy*, Appl. Phys. A **66**, S367–S370 (1998)
23. J.P. Fillard: *Near-Field Optics and Nanoscopy* (World Scientific, Singapore 1996) ISBN 981-02-2349-8
24. I. Banno, T. Inoue, H. Hori: *An Ampere-Like Law for Displacement Vector Field and Near-Field Optical Microscopy*, Opt. Rev. **3** (6B), 454–457 (1996)
25. E. Paule, P. Reineker: *Near-Field Optical Microscopy: A Numerical Study of Resolution and Contrast*, J. of Luminescence **76–77**, 299–302 (1998)
26. A.J. Ward, J.B. Pendry: *The Theory of SNOM: A Novel Approach*, J. of Mod. Opt. **44** (9), 1703–1714 (1997)
27. K. Cho, K. Arima, Y. Ohfuti: *Theory of Resonant SNOM (Scanning Near-Field Optical Microscopy): Breakdown of the Electric Dipole Selection Rule in the Reflection Mode*, Surf. Sci. **363**, 378–384 (1996)
28. S. Li, W. Liu, X. Li: *Restoration Mode Scanning Near-Field Optical Microscope Image*, Proc. SPIE BIOS 1999 **3607**, 146–157 (1999)
29. B. Knoll, F. Keilmann: *Electromagnetic Fields in the Cutoff Regime of Tapered Metallic Waveguides*, Opt. Comm. **162**, 177–181 (1999)
30. H. Muramatsu, K. Honma, N. Chiba, K. Nakajima, T. Ataka, S. Ohta, A. Kusumi, M. Fujihira: *Scanning Near-Field Optic/Atomic Force Microscopy in Liquids*, Thin Solid Films **273**, 335–338 (1996)
31. U. Durig: *Interaction Sensing in Dynamic Force Microscopy*, New J. of Phys. **2**, 5.1–5.12 (2000)
32. S. Heisig, E. Oesterschulze: *Gallium Arsenide Probes for Scanning Near-Field Probe Microscopy*, Appl. Phys. A **66**, S385–S390 (1998)
33. S. Heisig, A. Leyk, H.U. Danzebrink, W. Mertin, S. Munster, E. Oesterschulze: *Monolithic Gallium Arsenide Cantilever for Scanning Near-Field Microscopy*, Ultramicroscopy **71**, 99–105 (1998)
34. G. Schurmann, U. Staufer, P.F. Indermuhle, N.F. de Rooij: *Micromachined SPM Probes with sub-100nm Features at Tip Apex*, Surf. and Interf. Anal. **27**, 299–301 (1998)
35. S. Werner, C. Mihalcea, O. Rudow, E. Oesterschulze: *Cantilever Probes with Aperture Tips for Polarization-Sensitive Scanning Near-Field Optical Microscopy*, Appl. Phys. A **66**, S367–S370 (1998)
36. S. Munster, D. Mihalcea, S. Werner, W. Scholz, E. Oesterschulze: *Novel Micromachined Cantilever Sensors for Scanning Near-Field Optical Microscopy*, J. of Microscopy **186** (1), 17–22 (1997)
37. N.F. van Hulst, O.F.J. Noordman, M.H.P. Moers, R.G. Tack, F.B. Segerink, B. Bolger: *Near-Field Optical Microscope Using a Silicon Nitride Probe*, Appl. Phys. Lett. **65** (5), 461–463 (1993)
38. J.P. Cleveland, D. Bocek, S. Manne, P.K. Hansma: *A Nondestructive Method for Determining the Spring Constant of Cantilevers for Scanning Force Microscopy*, Rev. Sci. Inst. **64** (2), 403–405 (1993)
39. H. Muramatsu, N. Yamamoto, N. Chiba, K. Honma, T. Ataki, M. Shigeno, H. Monobe, M. Fujihira: *Multifunctional SNOM/AFM Probe with Accurately Controlled Low Spring Constant*, Ultramicroscopy **71**, 73–79 (1998)

40. J.A. Veerman, L. Kuipers, A.M. Otter, N.F. van Hulst: *High Definition Aperture Probes for Near-Field Optical Microscopy Fabricated by Focused Ion Beam Milling*, Appl. Phys. Lett. **72** (24), 3115 (1998)
41. N. Chiba, H. Muramatsu, T. Ataka, M. Fujihira: *Observation of Topography and Optical Image of Optical Fiber End by Atomic Force Mode Scanning Near-Field Optical Microscope*, Jpn. J. Appl. Phys. **34**, 321–324 (1995)
42. B. Hecht, H. Heinzelman, D.W. Pohl, L. Novotny: *Tunnel Near-Field Optical Microscopy: TNOM-2*, Ultramicroscopy **61**, 99–104 (1995)
43. A.J. Meixner, H. Kneppe: *Scanning Near-Field Optical Microscopy in Cell Biology and Microbiology*, Cellular and Molecular Biology **44** (5), 673–688 (1998)
44. V. Subramaniam, T.M. Jovin, A.K. Kirch: *Cell Biological Applications of Scanning Near-Field Optical Microscopy (SNOM)*, Cellular and Molecular Biology **44** (5), 689–700 (1998)
45. A. Lewis, N. Ben-Ami, A. Radko, D. Palanker, K. Lieberman: *Near-Field Scanning Optical Microscopy in Cell Biology*, Cell Biology **9**, 70–73 (1999)
46. S. Wang, J.M. Siqueiros: *Influence of the Sample–Probe Coupling on the Resolution of Transmitted Near-Field Optical Image*, Optics and Lasers in Engineering **31**, 517–527 (1999)
47. U. Durig, F. Rohner, D.W. Pohl: *Near-Field Optical Scanning Microscopy*, J. of Appl. Phys. **59**, 3318–3327 (1986)
48. E. Betzig, R.J. Chichester: *Single Molecules Observed by Near-Field Scanning Optical Microscopy*, Science **262**, 1422–1425 (1993)
49. O. Hollricher, O. Marti, R. Brunner: *Piezoelectrical Shear-Force Distance Control in Near-Field Optical Microscopy for Biological Applications*, Ultramicroscopy **71**, 143–147 (1998)
50. N.F. van Hulst, M.H.P. Moers, O.F.J. Noordman, R.G. Tack, F.B. Segerink, B. Bolger: *Near-Field Optical Microscope Using a Silicon Nitride Probe*, Appl. Phys. Lett. **62**, 461–463 (1993)
51. C. Wright-Smith, C.M. Smith: *Atomic Force Microscopy*, The Scientist **15** (2), 23 (22 January 2001)
52. P.K. Hansma, J.P. Cleveland, M. Radmacher, D.A. Walters, P.E. Hiliner, M. Bezanilla, M. Fritz, D. Vie, H.G. Hansma: *Tapping Mode Atomic Force Microscopy in Liquids*, Appl. Phys. Lett. **64**, 1738–1740 (1994)
53. A.J. Constant, K.O. Van der Werf, J. Putman, B.G. De Grooth, N.F. Van Hulst, J.N Greve: *Tapping Mode Atomic Force Microscopy in Liquid*, Appl. Phys. Lett. **64** (18), 2454–2456 (1994)
54. M. Dreier, T. Richmond, D. Anselmetti, U. Dammer, H.-J. Guntherodt: *Dynamic Force Microscopy in Liquids*, J. of Appl. Phys. **76** (9), 5095–5098 (1994)
55. B. Perner, L. Wolleber, M. Hausmann, A. Rapp, S. Monajembashi, K.-O. Greulich: *Scanning Near-Field Optical Microscopy of Cell Surfaces after Structure Conserving Air Drying*, Proc. SPIE BIOS 2000 (January 2000)
56. J.A. DeRose, J.P. Revel: *Studying the Surface of Soft Materials (Live Cells) at High Resolution by Scanning Probe Microscopy: Challenges Faced*, Thin Solid Films **331**, 194–202 (1998)
57. H. Muramatsu, T. Umemoto, N. Chiba, K. Honma, K. Nakajima, T. Ataka, S. Ohta, A. Kusumi, M. Fujihira: *Development of Near-Field Optic/Atomic Force Microscope for Biological Materials in Aqueous Solution*, Ultramicroscopy **61**, 265–269 (1995)
58. H. Muramatsu, T. Ataka, N. Chiba, H. Monobe, M. Fujihira: *Scanning Near-Field Optic/Atomic Microscopy*, Ultramicroscopy **57**, 141–146 (1995)

59. M. Fujihira, H. Muramatsub, T. Ataka, H. Monobea: *Measurements of Lateral Distribution of Fluorescence Intensities and Fluorescence Spectra of Microareas by a Combined SNOM and AFM*, Ultramicroscopy **57** (2–3), 118–123 (1995)
60. H. Muramatsu, N. Chiba, K. Homma, K. Nakajima, T. Ataka, S. Ohta, A. Kusumi, M. Fujihira: *Near-Field Optical Microscopy in Liquids*, Appl. Phys. Lett. **66**, 3245–3247 (1995)
61. N. Chiba, H. Muramatsu, T. Ataka, M. Fujihira: *Observation of Topography and Optical Image of Optical Fiber End by Atomic Force Mode Scanning Near-Field Optical Microscope*, Jpn. J. Appl. Phys. **34**, 321–324 (1995)
62. T. Ataka, K. Nakajima, H. Muramatsu, N. Chiba, K. Honma, M. Fujihira: *Design and Application of Scanning Near-Field Optical/Atomic Force Microscopy*, Thin Solid Films **273**, 154–160 (1996)
63. J.D. Bui, H.J. Lou, T. Zelles, V.L. Gallion, M.I. Philips, W. Tan: *Probing Intracellular Dynamics in Living Cells with Near-Field Optics*, J. of Neuroscience Methods **89**, 9–15 (1999)
64. S. Iwabuchi, Y. Morita, A. Hashigasako, T. Sakaguchi, Y. Murakami, K. Yokoyama, E. Tamiya: *Advanced Imaging for DNA Analysis Based on Scanning Near-Field Optical/Atomic-Force Microscopy (SNOAM)*, Proc. SPIE BIOS 1999 **3607**, 102–107 (January 1999)
65. S. Iwabuchi, N. Chiba, H. Muramatsu, Y. Kinjo, Y. Murakami, T. Sakaguchi, K. Yokoyama, E. Tamiya: *Simultaneous Detection of Near-Field Topographic and Fluorescence Imagings of Human Chromosomes via Scanning Near-Field Optical/Atomic-Force Microscopy (SNOAM)*, Nucleic Acids Res. **25** (8), 1662–1663 (1997)
66. M. Chalfie, G. Euskirchen, Y. Tu, W.W. Ward, D.C. Prasher: *Green Fluorescent Protein as a Marker for Gene Expression*, Science **263** (5148), 802–805 (1994)
67. H.H. Gerdes, C. Kaether: *Green Fluorescent Protein: Applications in Cell Biology*, FEBS Lett. **389** (1), 44–47 (1996)
68. T. Schmidt, W. Baumgartner, G.J. Schutz, H.J. Gruber, H. Schindler: *Imaging of Single Molecule Diffusion*, PNAS **93** (7), 2926–2929 (1996)
69. T. Funatsu, M. Tokunaga, Y. Harada, K. Saito, T. Yanagida: *Imaging of Single Fluorescent Molecules and Individual ATP Turnovers by Single Myosin Molecules in Aqueous Solution*, Nature **374** (6522), 496 (1995)
70. E. Tamiya, N. Nagatami, S. Iwabuchi, Y. Murakami, T. Sakaguchi, K. Yokoyama: *Simultaneous Topographic and Fluorescence Imaging of Recombinant Bacterial Cells Containing a Green Fluorescent Protein Gene Detected by a Scanning Near-Field Optical/Atomic Force Microscope*, Anal. Chem. **69**, 3697–3701 (1997)
71. H. Muramatsu, T. Ataka, N. Chiba, S. Iwabuchi, N. Nagatani, E. Tamiya, M. Fujihira: *Scanning Near-Field Optical/Atomic Force Microscopy for Fluorescence Imaging and Spectroscopy of Biomaterials in Air and Liquid: Observation of Recombinant Escherichia Coli with Gene Coding to Green Fluorescent Protein*, Opt. Rev. **3** (6B), 470–474 (1996)
72. D.C. Prasher, W.W. Ward, V.K. Eckenrode, F.G. Prendergast, M.J. Cormier: *Primary Structure of the Aequorea Victoria Green-Fluorescent Protein*, Gene **111** (2), 229–233 (1992)
73. R. Heim, R.Y. Tsien, D.C. Prasher: *Wavelength Mutation and Post-Translational Autooxidation of Green Fluorescent Protein*, PNAS **91** (26), 12501 (1994)

74. T.W. Berger, R.J. Sclabassi, G. Chaivet: *A Biologically Based Model of Function Properties of the Hippocampus*, Neural Networks **7** (6–7), 1031–1064 (1994)
75. T.V.P. Bliss, G.L. Collingridge: *A Synaptic Model of Memory: Long Term Potentiation in the Hippocampus*, Nature **361**, 31–39 (1993)
76. M.D. Ehlers, L.-F. Lau, A.L. Mammen, R.L. Hungair: *Synaptic Targeting of Glutamate Receptors*, Current Opinion in Cell Biology **8**, 484–489 (1996)
77. R. Tokioka, K. Kiyosue, A. Matsuo, M. Kasai, T. Taguchi: *Synapse Formation in Dissociated Cell Cultures of Embryonic Chick Cerebral Neurons*, Brain Res. Dev. Brain Res. **74** (1), 145–150 (1993)
78. P. Degenaar, Y. Murakami, K. Yokoyama, E. Tamiya: *Near-Field Imaging of Neurotransmitter Release and Uptake in Patterned Neuron Networks*, Proc. SPIE Progress in Biomedical Optics and Imaging II **1**, (16), 189–198 (2000)
79. T. Niwa, K. Kato, Y. Mitsuoka, S. Ichihara, N. Chiba, M. Shin-Ogi, K. Nakajima, H. Muramatsu, T. Sakuhara: *Optical Microcantilever Consisting of Channel Waveguide for Scanning Near-Field Optical Microscopy Controlled by Atomic Force*, J. Microscopy **194** (2/3), 388–392 (1999)
80. A. Lewis, N. Ben-Ami, K. Lieberman, G. Fish, E. Khachatryan, U. Ben-Ami, S. Shalom: *New Design and Imaging Concepts in NSOM*, Ultramicroscopy **61**, 215–220 (1995)
81. A.G.T. Ruiter, A. Jalocha, M.H.P. Moers, N.F. van Hulst: *Development of an Integrated NSOM Probe*, Ultramicroscopy **61**, 139–143 (1995)
82. J. Kim, J.T. Boyd, D.E. Pride, H.E. Jackson: *Spectrally-Resolved Near-Field Investigation of Proton Implanted Vertical Cavity Surface Emitting Lasers*, Appl. Phys. Lett. **72** (24), 3112–3114 (1998)
83. U. Schwarz, D. Courion, M.L. Berthie, H. Bielefeldt: *Simple Refelction Scanning Near-Field Optical Microscope Using the Back-Reflected Light Inside the Laser Cavity as Detection Mode*, Opt. Comm. **134**, 301–309 (1997)
84. A. Jenei, V. Subramaniam, A.K. Kirch, D.J. Arndt-Jovin, T.M. Jovin: *Picosecond Multiphoton Scanning Near-field Optical Microscopy*, Biophys. J. **76**, 1092–1110 (1999)
85. A.K. Kirsch, G. Striker, V. Subramaniam, C. Schnetter, D.J. Arndt-Jovin, T.M. Jovin: *Continuous Wave Two-Photon Scanning Near-Field Optical Microscopy*, Biophys. J. **75**, 1513–1521 (1998)
86. H. Yamada, S. Watanabe, H. Itoh, K. Kobayashi, K. Masushige: *Scanning Near-Field Optical Microscopy Using Piezoelectric Cantilevers*, Surf. and Interf. Anal. **27**, 503–506 (1999)
87. A.K. Kirsch, T.M. Jovin, C.K. Meyer: *Shear Force Imaging of DNA in Near-Field Scanning Optical Microscope (NSOM)*, J. of Microscopy **185** (3), 396–401 (1997)
88. H. Muramatsu, N. Yamamoto, K. Honma, J. Wang, K. Sakata-Sogawa, N. Shimamoto: *Imaging of DNA Molecules by Scanning Near-Field Microscopy*, Mat. Sci. and Eng. C **12**, 29–32 (2000)
89. H. Furukawa, S. Kawata: *Local Field Enhancement with an Apertureless Near-Field Microscope Probe*, Opt. Comm. **148**, 221–224 (1998)
90. S.A. Vickery, R.C. Dunn: *Scanning Near-Field Fluorescence Resonance Energy Transfer Microscopy*, Biophys. J. **76**, 1812–1818 (1999)
91. D.T. Schaafsma: *Single Mode Chalcogenide Fiber Infrared SNOM Probes*, Ultramicroscopy **77**, 77–81 (1999)

92. Y.-W. Shi, S. Mori, Y. Abe, Y. Matsunura, M. Miyagi: *Polymer-Coated Hollow Fibers for Medical Infrared Lasers: Fabrication, Transmission, and End-Sealing*, Proc. Photonics West 2000 c **1** (2000)
93. M.K. Hong, N.V. Dokholyan, A.G. Jeung, T.I. Smith, H.A. Schwettaman, P. Huie, S. Erramilli: *Imaging Living Cells with a Scanning Near-Field Infrared Microscope Based on a Free Electron Laser*, NIM B **144**, 246–255 (1998)
94. T.A. Klar, M. Dyba, S. Jakobs, A. Egner, S.W. Hell: *Fluorescence Microscopy with Diffraction Resolution Barrier Broken by Stimulated Emission*, PNAS **97** (15), 8206–8210 (2000)

5 Atomic Force Microscopy for Imaging Living Organisms: From DNA to Cell Motion

Tatsuo Ushiki

Scanning probe microscopes (SPM) form a new family of microscopes developed from the invention of the scanning tunneling microscope (STM) by Binnig and Rohrer in 1981 [1]. These microscopes have no lens but simply scan a sharp probing tip over a sample surface, providing high resolution information about the surface characteristics of solid samples. Among them, the atomic force microscope (AFM) invented by Binnig, Quate and Gerber in 1986 [2] has been widely applied to the study of surface structures, especially in the field of material science [3]. This is because the AFM can create topographic images of sample surfaces from the micrometer scale to the atomic scale. This microscope has always been expected to become an essential tool for the study of biological structures for the following reasons:

- The AFM enables direct observation of non-conductive samples without metal coating or any conductive treatment, in contrast to scanning electron microscopy.
- The AFM has the advantage that it creates images not only in vacuum but also in an air or liquid environment, indicating that functional information can be expected concerning living cells.
- The AFM provides quantitative information on the sample height, in contrast with scanning electron microscope images.

A great many studies have thus been published on applications of AFM to biological studies during the past decade [4–6]. We have also observed various kinds of biological sample from macromolecules to living cells by AFM [7–9]. In this chapter, we briefly introduce the principles underlying AFM and present our own results in the biomedical field, focusing on the current status and future of biological AFM.

5.1 Principles of Atomic Force Microscopy

The AFM is roughly composed of the microscope unit, control unit, and computer (Fig. 5.1). In the microscope unit, the probe is attached to the end of a cantilever. When the probe approaches the sample surface, a tiny interaction force occurs between the probe and sample, and this influences the position of the cantilever. The cantilever position is usually measured by the optical lever method. Using an appropriate feedback system in the control unit, the distance between the probe and sample is accurately regulated by

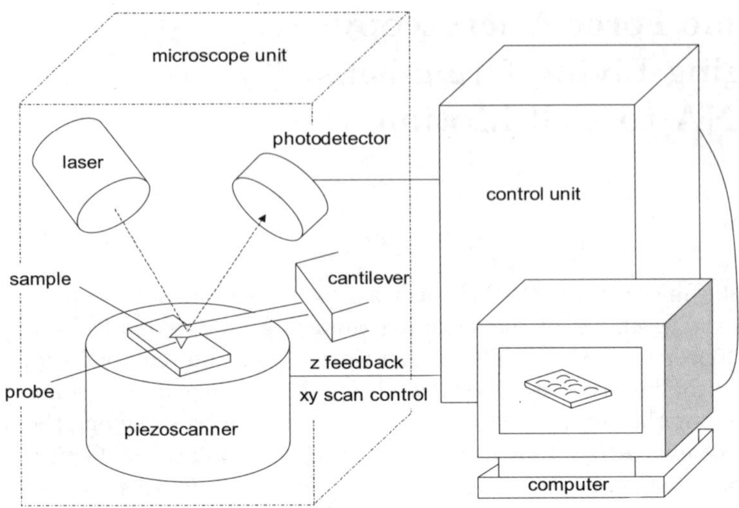

Fig. 5.1. The schematic illustration of the atomic force microscope

changing the height of the sample. A piezoelectronic xyz-scanner is usually used both to control the vertical position of the sample (z) and to raster-scan the sample in the xy-plane.

There are two different operating modes in AFM: a contact mode and a dynamic force mode. In contact mode AFM, the tip is simply dragged across the sample surface whilst measuring the deformation of the cantilever. Dynamic force mode AFM, on the other hand, scans the sample surface with a vibrating cantilever. In this mode, the probe–sample interaction force is controlled by alterations in the resonance frequency of the cantilever, since the resonance frequency shifts to reflect changes in probe–sample distance. The vertical interaction force in dynamic force mode AFM can be smaller (10^{-9}–10^{-10} N) than in contact mode AFM (about 10^{-8} N) and is suitable for imaging very soft samples.

AFM images are thus created either by measuring the cantilever deflection (in contact mode AFM) or the resonance frequency (in dynamic force AFM). Images produced by the feedback voltage are called constant force images (or height images). These reflect the topographical height of the sample. On the other hand, images constituted by changes in the cantilever deflection (in contact mode AFM) or the oscillation frequency (in dynamic force mode AFM) are called variable deflection images. These are suitable for visualizing the details of sharp contour changes in the sample.

There are different graphic modes for computer display. The most popular mode is the gradation mode, which represents the sample height or cantilever deflection by color gradations or a gray scale. In constant force images, the shading mode displayed by an illuminated three-dimensional rendering is also used to represent surface details of specimens.

5.2 Applications in Biology

5.2.1 Deoxyribonucleic Acid (DNA) and Chromosomes

Deoxyribonucleic acid (DNA) is a polymer of nucleotides. Visualization of DNA by AFM has attracted many researchers, since the structure of DNA is simple and is considered to be suitable for AFM study. The most common method for DNA preparation is to deposit DNA solution on the substrate and then evaporate the solvent. Although various materials such as sapphire and graphite have been reported [10], freshly cleaved mica is widely used as the substrate. Previous studies have reported that DNA molecules do not bind strongly to mica and are usually moved or swept away by the scanning tip, because freshly cleaved mica and DNA are both negatively charged. Therefore, the mica surface is often treated with Mg^{++} and a variety of other divalent and trivalent cations to prepare for AFM observation of DNA [11]. On the other hand, we showed that DNA molecules dissolved in distilled water also bind consistently and strongly enough to freshly cleaved mica to be able to image with dynamic force mode AFM (Fig. 5.2).

AFM images of DNA are comparable to those taken by a rotary shadowing technique for transmission electron microscopy (TEM). However, AFM has the further advantage of directly observing the molecule without staining or shadowing. AFM images also have the advantage of containing quantitative information on the sample height. The height of DNA images is usually about 1 nm or less, while the image width varies typically over the range 8–20 nm. The reason why the width of DNA is much larger than its height is explained by the fact that the probing tip measures about 10 nm across, which is much larger than the diameter of the DNA molecule. The application of carbon

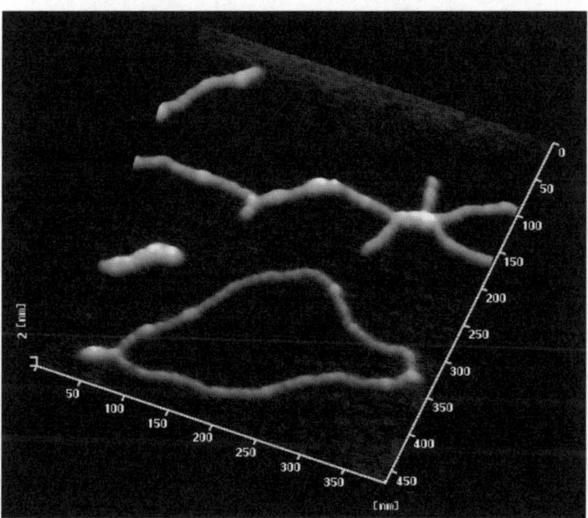

Fig. 5.2. Circular plasmid DNA (pUC 18) observed in a dynamic force mode AFM in air. Note the supercoiled DNA (*arrows*)

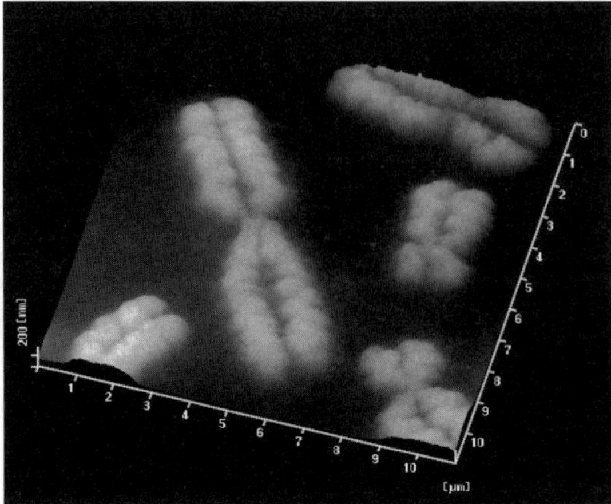

Fig. 5.3. Human metaphase chromosomes observed in a dynamic force mode AFM in air. Note ridges and grooves on the chromosomes (this micrograph was taken by Dr. Osamu Hoshi, Niigata University)

nanotube probes is effective for the AFM analysis of the detailed structure of DNA [12].

Chromosomes in eukaryotic cells are an assembly of chromatin fibers which consist of DNA and proteins. Chromosomes are usually prepared in accordance with the standard method for light microscopy [13]. Briefly, chromosome spreads are made by dropping a suspension of peripheral blood cultures (which are treated with a hypotonic solution of potassium chloride and fixed with a mixture of methanol and acetic acid). Metaphase chromosomes split into separate chromatids (Fig. 5.3). The width of the chromatids is about 1 μm. The average height of the chromatids in this image lies in the range 300–350 nm when the specimens are critical point-dried. The chromosomes often show circumferential grooves and ridges on the surface of the chromatids. These structures are similar to the G-banding pattern, which has been recognized in light microscopy by a Giemsa staining method of trypsinated chromosomes.

With observation at a high magnification, the chromosome surface shows a granular appearance on a 50–100 nm scale. In trypsinated chromosomes, fiber-like or coiled protrusions, about 50–100 nm in diameter, are clearly revealed by AFM. These structures are probably related to higher level folding of chromatin fibers, a subject currently under investigation [14,15].

5.2.2 Collagen Molecules and Collagen Fibrils

Collagen type I molecules are chemically formed by a helical coil of three polypeptide chains. Each collagen molecule is observed as a twisted thread by AFM (Fig. 5.4). The measured length of the molecules is concentrated mainly

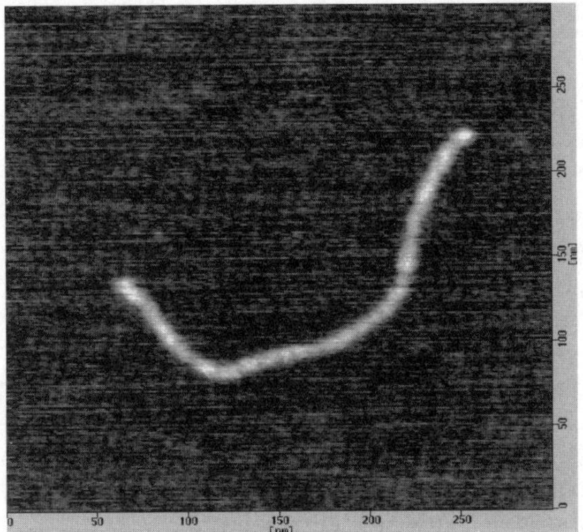

Fig. 5.4. Collagen type I molecule observed in a dynamic force mode AFM in air. An *arrow* indicates a depression on the molecule

between 280 and 310 nm, and the height and width are 0.5–1 nm and 6–10 nm, respectively [16]. At high magnification, the molecules have an uneven surface with bulges and depressions in several locations. Globular bulges are usually present at both ends of the molecules, while a prominent depression is often found at a point about 70 nm from one end of the molecules. A characteristic bending of the molecules is sometimes found at the opposite end.

Collagen fibrils imaged by AFM are semi-cylindrical in profile, about 100 nm in width, and 35–40 nm in height (Fig. 5.4) [7,17–19]. The reason why the width of fibrils is imaged about three times larger than their height is explained by the pyramidal shape of the probe [19]. Almost all fibrils examined by AFM showed a repeating pattern of grooves and ridges with a period of 60–70 nm. One of the advantages of AFM observation is that the profile of collagen fibrils can easily be obtained from the constant force images. The difference in height between the grooves and the ridges was 3–5 nm in the air-dried subcutaneous collagen fibrils. In addition, the subcutaneous fibrils often exhibit a shallow (less than 1 nm) depression in the middle of the ridges, and a small (less than 1 nm) elevation in the middle of the grooves [7]. A similar finding by AFM has been reported in collagen fibrils from the rat tail tendon [20], although the meaning of the depression and elevation is considered hard to explain in these air-dried specimens. AFM observation of collagen fibrils in physiological conditions (i.e., in saline solution) is eagerly awaited.

5.2.3 Tissue Sections

In order to analyze structures of cells and tissues in situ with an AFM, we need to prepare relatively flat sectioned or fractured samples because of the narrow z-range of the scanner. In fact it is usually limited to about 1 μm in the 20 μm scanner and 10 μm in the 100 μm scanner. Embedment-free tissue sections introduced in our previous paper are considered amongst the favorable samples for AFM observation (Fig. 5.5) [21]. Several investigators, on the other hand, have attempted to observe the surface of ultrathin sections of LR white or epoxy resin with an AFM [22,23]. It is true that the contours of cellular structures such as unit membranes and chromatin blocks are traceable even in the AFM images of the resin sections. However, these cellular images are merely produced by depressions or elevations in the resin, probably due to uneven cleavage at the site of cellular structures during ultrathin sectioning. On the other hand, AFM of embedment-free tissue sections is effective for direct observation of intracellular structures at high resolution.

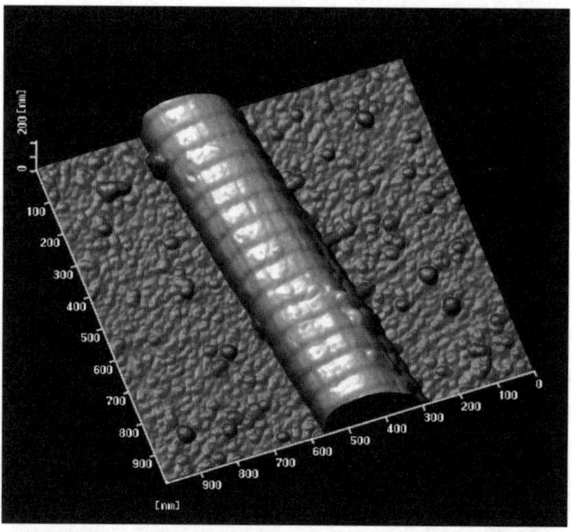

Fig. 5.5. AFM of a collagen fibril measured in a dynamic force mode in air. This fibril was isolated from the bovine sclera and critical point dried. Periodical grooves and ridges are obvious on the fibril

5.2.4 Living Cells and Their Movement

Because the AFM can be operated in liquid conditions, the surface topography of living cultured cells in a liquid environment can be clearly visualized. As noted in our previous papers, flat cells attached firmly to the substrate (i.e., a glass cover slip) are suitable for AFM imaging [7,24]. Spherical or

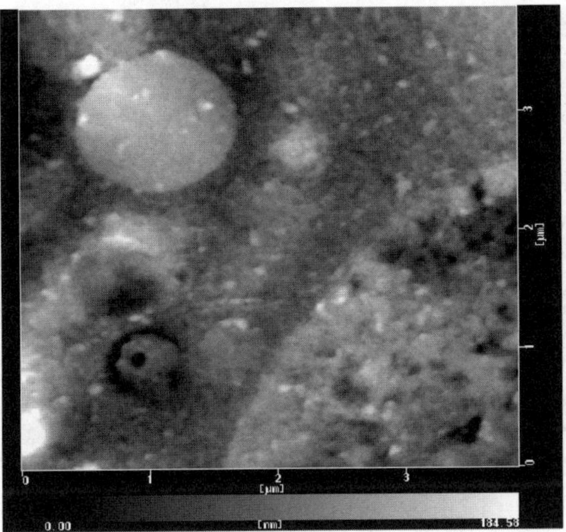

Fig. 5.6. AFM of the surface of the embedment-free section of a mouse pancreatic acinar cell

domed cells are easily detached from the glass surface during scanning, probably due to the lateral scanning force increasing at the steep slope of high samples. To minimize force-induced damage to the cell, it is also necessary to adjust the imaging force to the weakest value. Using contact mode AFM, cell processes such as lamellipodia are clearly observed at the cell margin (Fig. 5.6). For analysis of cell surface undulation, variable deflection mode images are useful, because these emphasize the contour of the cellular processes. Height images are useful for measuring the thickness of these processes. Contact mode AFM also provides information on the contour of cytoskeletal elements just beneath the cell membrane.

Since AFM can obtain images of living cells in fluids as above, AFM can be used to analyze dynamic events relating to living cells in real time. We have reported our attempt to collect sequential AFM images of living cultured cells [24,25]. Our previous studies revealed that the practical scan speed for observing living cells in liquids is under 20 μm/s in our instrument. Using this scan speed or less with the weakest imaging force, the movement of cell processes is successfully traced by contact mode AFM. This means that it took 2 min or more to obtain a single image with a 40 μm scan area, even when we used a 256×128 pixel size. For example, a series of AFM images with a 40 μm scan area were acquired at time intervals of 2–8 min. A fluid chamber system further enabled us to obtain AFM images of living cells over a period of one hour in perfectly adequate conditions [25].

To produce time-lapse movies, images were imported into the NIH Image program. Sequential images displayed by the same graphic mode were assembled into a movie using the NIH Image program and played back at $\sim 1\,000\times$

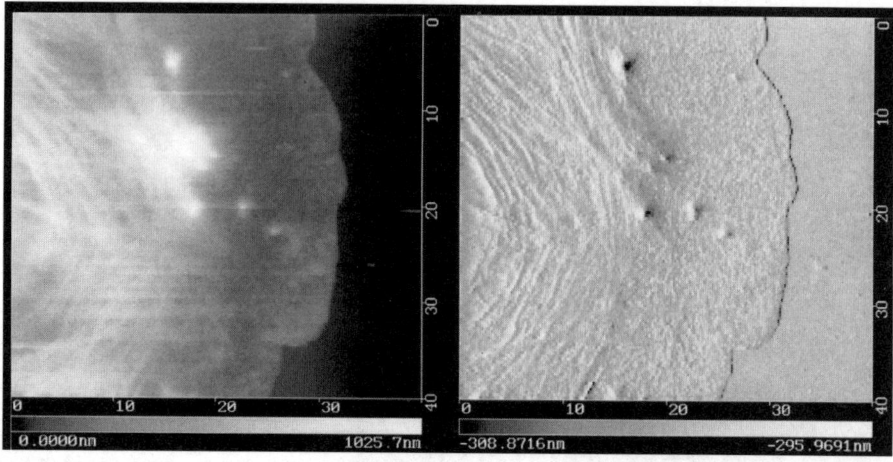

Fig. 5.7. AFM images of a part of the living cultured cell (human esophageal squqmous cell carcinoma cell, KESC2, C7 subclone) observed in culture medium. The height image (*left*) and variable deflection image (*right*) have simultaneously captured during scanning

real time on Macintosh computers. The realism of the movie records was useful in the high-resolution analysis of cellular and subcellular movements.

5.3 Other SPM Applications in Biology

Recent advances in scanning probe microscopes (SPM) have made it possible to collect topographic and other physiological information simultaneously from the same regions of samples. Viscoelastic images visualize differences in surface stiffness or elasticity of living culture cells [26–28]. On the other hand, scanning near-field optical/atomic force microscopy (SNOM/AFM) simultaneously provides high resolution topographic and optical images. We have applied SNOM/AFM to the observation of cultured cells and succeeded in obtaining fluorescence images of immunostained cytoskeletons simultaneously with topographic images of the same regions in a liquid environment [29]. Thus, combination of AFM and other SPM will provide us with further useful information concerning the structure and function of living cells and other biological samples [30].

5.4 Conclusion

The present chapter has been concerned with the application of AFM to biological studies from biomolecules to living cells. AFM instrumentation and techniques for preparing specimens for biological studies are still advancing, and we stand at the threshold of a new field of biology using AFM. AFM images provide a great deal of information which cannot be obtained by other types of microscopy such as electron microscopy. We thus believe that AFM is on the way to becoming an essential technique in the study of biological structures.

References

1. G. Binnig, H. Rohrer: *Scanning tunneling microscope*, Helv. Phys. Acta **55**, 726–735 (1982)
2. G. Binnig, C.F. Quate, C. Gerber: *Atomic force microscope*, Phys. Rev. Lett. **56**, 930–933 (1986)
3. R. Wiesendanger: *Scanning Probe Microscopy and Spectroscopy. Methods and Applications* (Cambridge University Press, Cambridge 1994)
4. R. Lal, S.A. John: *Biological applications of atomic force microscopy*, Am. J. Physiol. **266**, C1–C21 (1994)
5. A. Ikai: *STM and AFM of bio/organic molecules and structures*, Surf. Sci. Reports **26**, 261–332 (1996)
6. B.P. Jena, J.K.H. Hrber: *Atomic Force Microscopy in Cell Biology* (Academic Press, San Diego 2002)
7. T. Ushiki, J. Hitomi, S. Ogura, T. Umemoto, M. Shigeno: *Atomic force microscopy in histology and cytology*, Arch. Histol. Cytol. **59**, 421–431 (1996)
8. T. Ushiki, S. Ogura, J. Hitomoi, T. Umemoto, M. Shigeno: *Observation of cells and tissues by atomic force microscopy*, in *Recent Advances in Microscopy of Cells, Tissues and Organs*, ed. by P.M. Motta (Antonio Delfino Editore, Rome 1997) pp. 59–63
9. T. Ushiki: *Atomic force microscopy and its related techniques in biomedicine*, It. J. Anat. Embryol **106 (suppl. 1)**, 3–8 (2001)
10. K. Yoshida, M. Yoshimoto, K. Sasaki, T. Ohnishi, T. Ushiki, J. Hitomi, S. Yamamoto, M. Shigeno: *Fabrication of a new substraste for atomic force microscopic observation of DNA molecules using the ultrasmooth sapphire plate*, Biophys. J. **74**, 1654–1657 (1998)
11. Z. Shao, J. Yang: *Progress in high resolution atomic force microscopy in biology*, Quart. Rev. Biophys. **28**, 195–251 (1995)
12. K.I. Hohmura, Y. Itokazu, S.H. Yoshimura, G. Mizuguchi, Y.S. Masamura, K. Takeyasu, Y. Shiomi, T. Tsurimoto, H. Nishijima, S. Akita, Y. Nakayama: *Atomic force microscopy with carbon nanotube probe resolves the subunit organization of protein complexes*, J. Electron Microsc. **49**, 415–421 (2000)
13. D.E. Rooney, B.H. Czepulkowski: *Human Chromosome Preparation. Essential Techniques* (John Wiley and Sons, Chichester 1997)
14. O. Hoshi, T. Ushiki: *Three-dimensional structure of G-banded human metaphase chromosomes observed by atomic force microscopy*, Arch. Histol. Cytol. **64**, 475–482 (2001)
15. T. Ushiki, O. Hoshi, K. Iwai, E. Kimura, M. Shigeno: *The structure of human metaphase chromosomes: Its histological perspective and new horizons by atomic force microscopy*, Arch. Histol. Cytol. **65**, 377–390 (2002)
16. S. Yamamoto, F. Nakamura, J. Hitomi, M, Shigeno, S. Sawaguchi, H. Abe, T. Ushiki: *Atomic force microscopy of intact and digested collagen molecules*, J. Electron Microsc. **49**, 473–481 (2000)

17. S. Yamamoto, J. Hitomi, M. Shigeno, S. Sawaguchi, H. Abe, T. Ushiki: *Atomic force microscopic studies of isolated collagen fibrils of the bovine cornea and sclera*, Arch. Histol. Cytol. **60**, 371–378 (1997)
18. S. Yamamoto. S., H. Hashizume, J. Hitomi, M. Shigeno, S. Sawaguchi, H. Abe, T. Ushiki: *The subfibrillar Arrangement of corneal and scleral collagen fibrils as revealed by scanning electron and atomic force microscopy*, Arch. Histol. Cytol. **63**, 127–135 (2000)
19. S. Yamamoto, J. Hitomi, S. Sawaguchi, H. Abe, M. Shigeno, T. Ushiki: *Observation of human corneal and scleral collagen fibrils by atomic force microscopy*, Jpn. J. Ophthalmol. **46**, 496–501 (2002)
20. D.R. Baselt, J.P. Revel, J.D. Baldeschwieler: *Subfibrillar structure of type I collagen observed by atomic force microscopy*, Biophys. J. **65**, 2644–2655 (1993)
21. T. Ushiki, M. Shigeno, K. Abe: *Atomic force microscopy of embedment-free sections of cells and tissues*, Arch. Histol. Cytol. **57**, 427–432 (1994)
22. A. Yamamoto, Y. Tashiro: *Visualization by an atomic force microscope of the surface of ultra-thin sections of rat kidney and liver cells embedded in LR White*, J. Histochem. Cytochem. **42**, 1463–1470 (1994)
23. S. Yamashina, M. Shigeno: *Application of atomic force microscopy to ultrastructural and histochemical studies of fixed and embedded cells*, J. Electron Microsc. **44**, 462–466 (1995)
24. T. Ushiki, S. Yamamoto, J. Hitomi, S. Ogura, T. Umemoto, M. Shigeno: *Atomic force microscopy of living cells*, Jpn. J. Appl. Phys. **39**, 3761–3764 (2000)
25. T. Ushiki, J. Hitomi J, T. Umemoto, S. Yamamoto, H. Kanazawa, M. Shigeno: *Imaging of living cultured cells of an epithelial nature by atomic force microscopy*, Arch. Histol. Cytol. **62**, 47–55 (1999)
26. S. Sasaki, M. Morimoto, H. Haga, K. Kawabata, E. Ito, T. Ushiki, K. Abe, T. Sambongi: *Elastic properties of living fibroblasts as imaged using force modulation mode in atomic force microscopy*, Arch. Histol. Cytol. **61**, 57–63 (1998)
27. H. Haga, S. Sasaki, K. Kawabata, E. Ito, T. Ushiki, T. Sambongi: *Elasticity mapping of living fibroblasts by AFM and immunofluorescence observation of the cytoskeleton*, Ultramicroscopy **82**, 253–258 (2000)
28. H. Haga, M. Nagayama, K. Kawabata, E. Ito, T. Ushiki, T. Sambongi: *Time-lapse viscoelastic imaging of living fibroblasts using force modulation mode in AFM*, J. Electron Microsc. **49**, 473–481 (2000)
29. H. Muramatsu, N. Chiba, K. Nakajima, T. Ataka, M. Fujihira, J. Hitomi, T. Ushiki: *Fluorescence imaging and spectroscopy of biomaterials in air and liquid by SNOM/AFM*, Scanning Microsc. **10**, 975–982 (1996)
30. E. Kimura, J. Hitomi, T. Ushiki: *Scanning near field optical/ atomic force microscopy of bromodeoxyuridine-incorporated human chromosomes*, Arch. Histol. Cytol. **65**, 435–444 (2002)

6 Expanding the Field of Application of Scanning Probe Microscopy

Hideki Kawakatsu

In the present chapter, we focus on projects carried out to exploit the potential of scanning probe microscopy [1–14] in the field of nanotechnology. The studies carried out can be categorized as follows:

- Nanotribology:
  - detect the position of the tip end for tribological studies in scanning force microscopy,
  - map the lateral vibration amplitude of the tip in scanning force microscopy,
  - create a linear scale using a crystal as the scale reference.
- Control:
  - develop a control method for eliminating drift at the subatomic level.
- Fabrication:
  - use nanometric mechanical oscillators as detectors of mass and force,
  - fabricate micrometric parallel leaf springs,
  - fabricate millions of cantilevers on a centimeter square chip.
- Characterization:
  - measure three-dimensional nanometric objects with a scanning probe microscope.

Nanotribology. A novel instrument incorporating two optical levers and a rectangular cantilever was implemented [15–17]. The instrument is capable of detecting displacements of the tip apex relative to the cantilever base, thus enabling visulization of the trajectory with subatomic resolution. It was shown that the tip apex meanders in a way that depends on the crystalline orientation of the sample.

The lateral vibration amplitude of a tip in contact with a solid sample differs according to the sample and contact conditions. Even on cleaved graphite surfaces, a clear contrast can be seen. We introduce the instrumentation and discuss the contrast mechanism.

As an application of results obtained from our nanotribological studies, we have implemented a linear displacement encoder, using a crystal as the scale reference. By selecting the crystal and conditions of operation, we have demonstrated a count rate of up to 100 000 lattice/s. The linear encoder provides an accurate scale for comparing measurements of small objects.

Control. Active position control of a sample stage was implemented by keeping the stage in registry with a crystalline lattice [18–21]. This enables averaging of weak signals from samples without drift of the viewing area. It was also demonstrated that atomic step height can be measured with a crystalline lattice as scale reference [22].

Fabrication. In the study of scanning probe microscopy, there often arises a need to implement novel detectors to enable imaging of small signals or to enable signal acquisition of very fast phenomena. Section 6.3 focuses on some fabrication techniques developed in the laboratory. We have so far demonstrated the fabrication of nanometric oscillators [23–27] for use in atomic force microscopy [2], and micrometric parallel leaf springs to be used when precise linear motion is required. The fabrication techniques utilize various methods developed for the fabrication of quantum dots [28,29]. However, in this case, the structures are three-dimensional and take the form of a mechanical oscillator or a mass supported by parallel leaf springs. Due to the drastic reduction in size, unprecedented resolution of force and mass detection can be expected. Preliminary results show the great strength of these three-dimensional nanometric structures [26,27].

Millions of cantilevers have been fabricated on a centimeter square chip. Each cantilever measures 5 μm long and 50 nm thick. Due to the reduced size and thickness of the cantilevers, a natural frequency of 10 MHz was obtained while keeping the spring constant at 10 N/m. The cantilever array can be used for simultaneous imaging or lithography. We focus on the fabrication process and some preliminary results.

Characterization. We have developed various techniques for making three-dimensional nanometric objects. These will be used to implement novel imaging methods in scanning probe microscopy. The need has therefore arisen to characterize the objects or detectors that we have fabricated.

To answer this need, we use a multiprobe scanning probe microscope that mounts on a commercial scanning electron microscope [30]. The design allows reasonable freedom in the angle of observation with the SEM, as well as force measurement and other characterizations with the scanning probe method. Since the samples were all three-dimensional with high aspect ratio, the visual information from the SEM helped to position the probes in three-dimensional space.

6.1 Nanotribology

6.1.1 An AFM with Two Optical Levers for Detecting the Trajectory of the Tip Apex

An optical lever or an optical fiber interferometer, as depicted in Figs. 6.1a and b [5,11], are widely used in scanning force microscopy to detect the

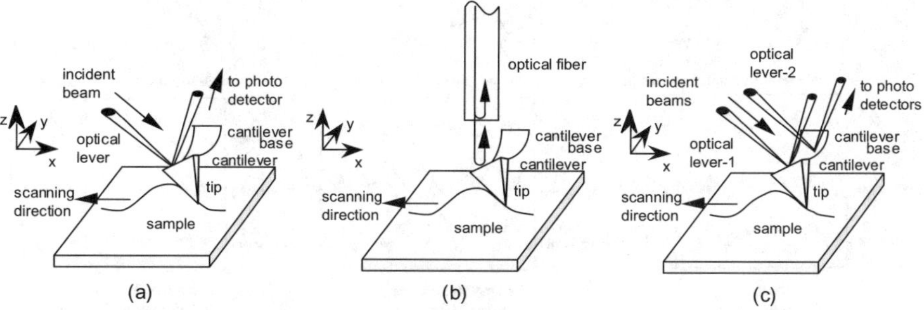

Fig. 6.1. Various optical methods for detecting cantilever displacement in atomic force microscopy: (**a**) optical lever detection, (**b**) optical fiber laser interferometry, and (**c**) dual optical lever detection

deformation of the cantilever. However, these detection schemes do not give enough information on the position of the tip apex relative to the cantilever base, leading to artefacts and difficulties in interpreting the behaviour of the tip during scanning. In order to detect deformations of the cantilever with multiple degrees of freedom, a scanning force microscope equipped with two optical levers was used, as depicted in Fig. 6.1c.

Figure 6.2 shows the deflection signals from the two optical levers when the sample was actuated by a known amount while the tip was in contact with the sample but without relative motion. Figures 6.2b and e are cases where the two laser beams are focused close to the end of the cantilever and close to the middle, respectively. When the sample was actuated by a small amount in the y and z directions so that the tip remained stuck to the sample, the polarity of the two deflection signals differed depending on the direction in which the sample was actuated. This means that by measuring the deflection and torsion at two different points on the cantilever, enough information is obtained to distinguish bending from buckling and thus detect the displacement of the tip end point relative to the base of the cantilever in the x, y and z directions.

The movement of the tip end point was monitored and plotted at regular time intervals when scanning mica in the repulsive mode. The plot revealed stick points reflecting the lattice structure as well as the trajectory of the tip. The sample was rotated with 30° increments to check the operation of the proposed detection system. The results are shown in Fig. 6.3. The tip followed an array of stick points but derailed every two or three lattices due to the misalignment of the lattice and the direction of the scan.

6.1.2 Mapping Lateral Tip Vibrations in Scanning Force Microscopy

As depicted in Fig. 6.4, the tip apex often vibrates at the torsional natural frequency of the cantilever, even when the tip comes into contact with the

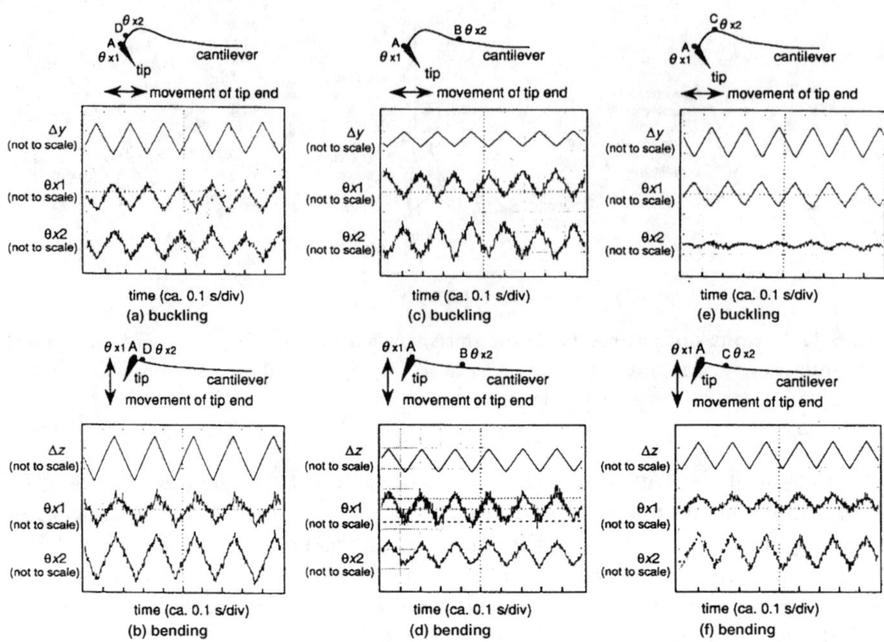

Fig. 6.2. Effect of changing the position of the focal points. The tip was brought into contact with the sample surface, and deflections θ_{x1} and θ_{x2} were measured for position modulation of the sample in the y and z directions. The amplitude of position modulation at a few hertz was of the order of 0.1 nm and remained small enough to ensure that the tip end point did not start to move relative to the sample. It can be assumed that the movement of the tip end point corresponds to the known movement of the sample. One focal point was always positioned close to the end of the cantilever (point A), whereas the other was positioned as follows: (**a**), (**b**) close to the end of the cantilever (point D); (**c**), (**d**) around the middle of the cantilever (point B); (**e**), (**f**) around one quarter of the way from the end of the cantilever (point C). (**a**), (**c**) and (**e**) correspond to position modulation in the y direction, and (**b**), (**d**) and (**f**) to position modulation in the z direction. It can be seen that the deflections θ_{x1} and θ_{x2} are of the same sign for (**a**), (**b**), (**d**), and (**f**). In (**c**), note that θ_{x1} and θ_{x2} are of different signs. In (**e**), the second focal point is chosen close to the node of deflection due to movement of the tip end point in the y direction

sample. This vibration is a relatively clear sinusoidal motion. By measuring the amplitude and mapping it while the sample is raster scanned in the x and y directions, we have demonstrated that contrast can often be seen in the amplitude image. The results of imaging by the proposed method are shown in Figs. 6.5 and 6.6 for silicon, silicon dioxide and graphite [31].

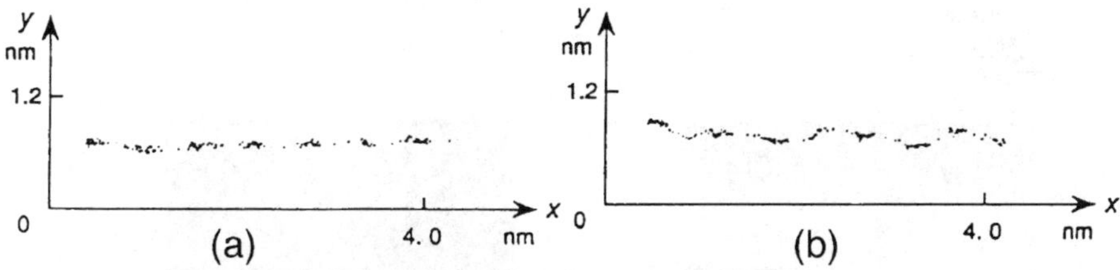

Fig. 6.3. Position of the tip end point plotted against xy coodinates for: (**a**), when the stick-points are aligned with the x axis, and (**b**) when the sample was rotated by approximately 15° about the z axis

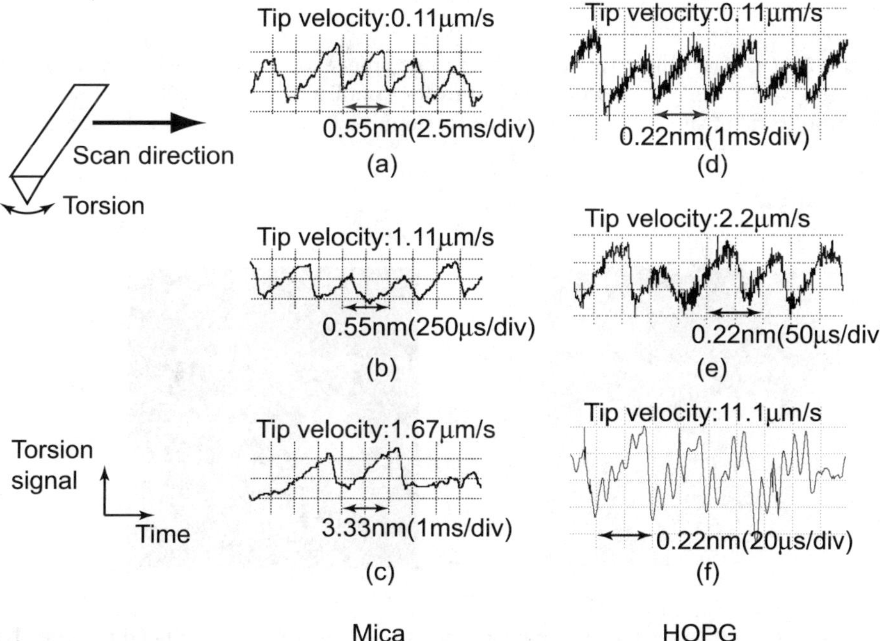

Fig. 6.4. Torsional signals of a friction force microscope for mica and graphite. The 450 kHz sinusoidal signal superposed onto the sawtooth stick–slip signal corresponds to the torsional natural frequency of the rectangular cantilever

6.1.3 Linear Scale Using a Crystal as Scale Reference

The crystalline lattice can be used as a scale reference. We have implemented various pieces of apparatus for comparing measurements in which the crystalline lattice is used as a reference for length and orientation [18–22]. We have concluded from our experiments that relative drift is the main factor degrading measurement accuracy, and that the best solution is to shorten the time required for measurement. As explained in Sect. 6.1.1, the tip end

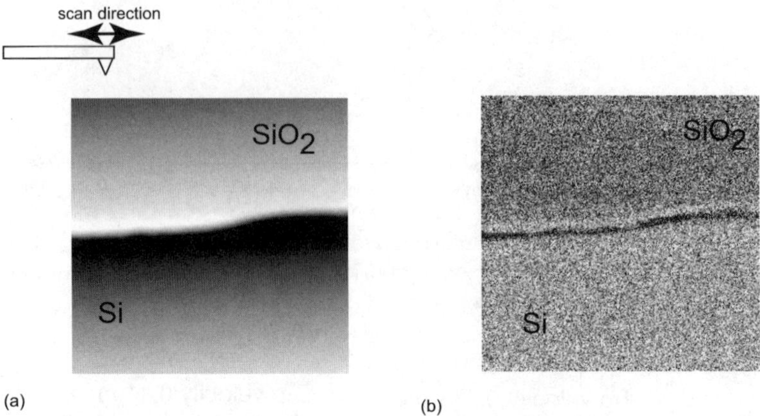

Fig. 6.5. (**a**) 1.1 µm × 1.1 µm topographic image of Si and SiO_2. Step height 60 nm. (**b**) 1.1 µm × 1.1 µm lateral vibration amplitude image (250 kHz) showing the amplitude of the twist signal at the torsional natural frequency of the cantilever

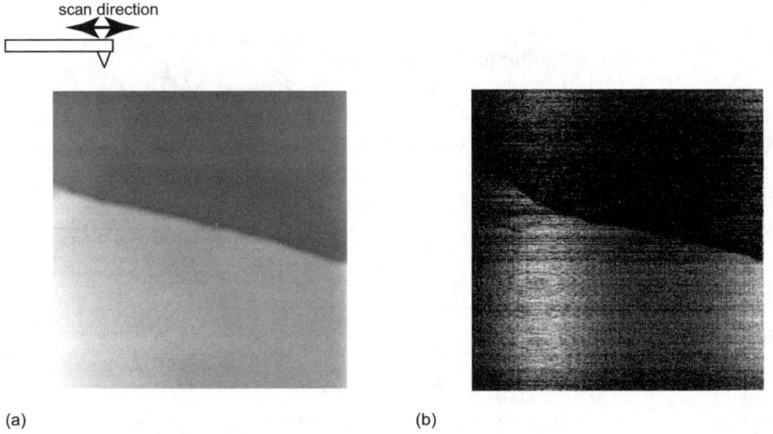

Fig. 6.6. (**a**) 1.1 µm × 1.1 µm topographic image of graphite. Step height 2 nm. (**b**) 1.9 µm × 1.9 µm lateral vibration amplitude image (250 kHz) showing the amplitude of the twist signal at the torsional natural frequency of the cantilever

of a scanning force microscope follows the array of stick points reflecting the lattice structure. By using this phenomenon, fast-counting of the number of atomic lattices can be implemented without actively controlling the probe to track an array of atoms [32]. When mica and graphite were used as the reference crystal, it was found that the maximum scanning velocity for mica without miscount was approximately 1 µm/s, whereas the value was 20 µm/s or higher for graphite. The latter corresponds to 80 000 lattice/s. We attribute the difference to the hydrophobic property of graphite, which helps to prevent the tip from being surrounded by a water droplet. At present, the natural frequency of the cantilever is 100 kHz and it determines the maximum scanning velocity without miscount. By using a compliant cantilever with high natural frequency in the 100 MHz range, we are planning to implement a linear encoder with scanning velocity of the order of 0.1 m/s.

6.2 Control

Implementing two-dimensional positioning control of a scanning tunneling microscope (STM) sample stage using a crystalline lattice as the scale reference, the mechanical movement of the stage can be regulated to the accuracy of the atomic spacing in a crystal [18–20]. Figure 6.7 depicts the control block diagram. A dither vibration of 70 pmp-p was applied to the tip in the x and y directions to implement differential imaging. Figure 6.8 shows the obtained differential signal. The zero crossings of the differential signal correspond to the local topographic maxima or minima. This means that either concave or convex surface stable scanning control can be implemented by feeding back the gradient signal to the x and y scanners, depending on the polarity of feedback.

Figure 6.9 shows the Lissajous figure of the tip position in x and y when the tip moves in registry with the lattice. Figures 6.9a and b correspond to cases where the step movement of the tip was set larger or smaller than the lattice spacing. This was carried out intentionally to confirm the operation of the proposed method for aligning the tip apex with the neighbouring atomic features. With a defect-free crystal, a regulated motion of the sample stage or the tip can thus be achieved without accumulative error. In a related study, a two-dimensional lateral tip positioning experiment in the sub-micrometer

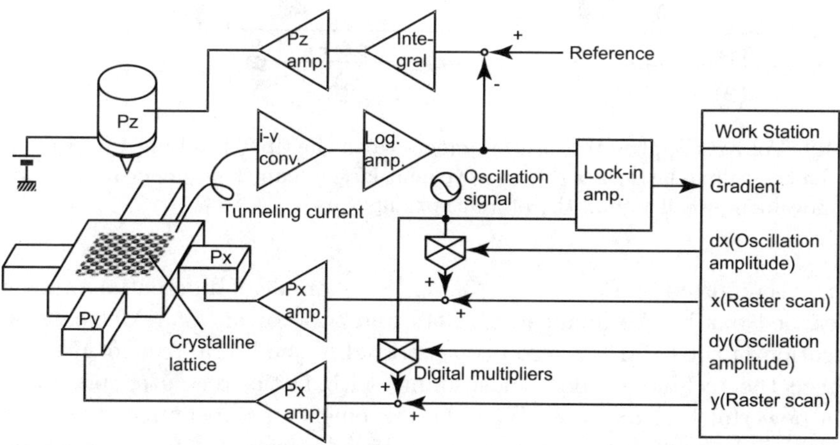

Fig. 6.7. Control block diagram for regulating the position of a sample stage by keeping the STM tip in registry with the atomic features of a crystalline lattice

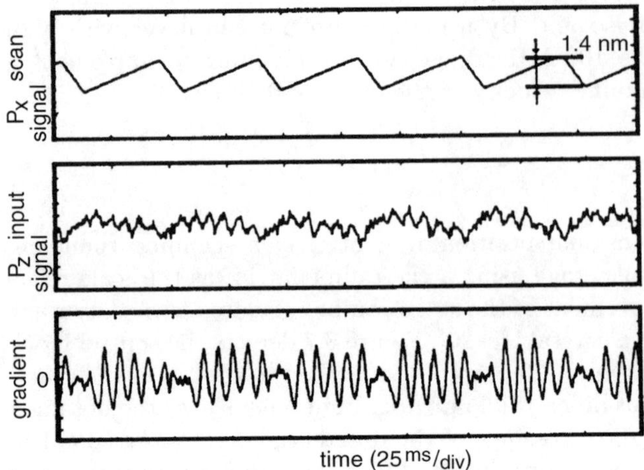

Fig. 6.8. *Top*: one line scan signal. *Centre*: *z* piezo voltage. *Bottom*: differential signal of the tunneling current. The zero crossings of the differential signal correspond to the local maxima and minima of the crystalline lattice topography

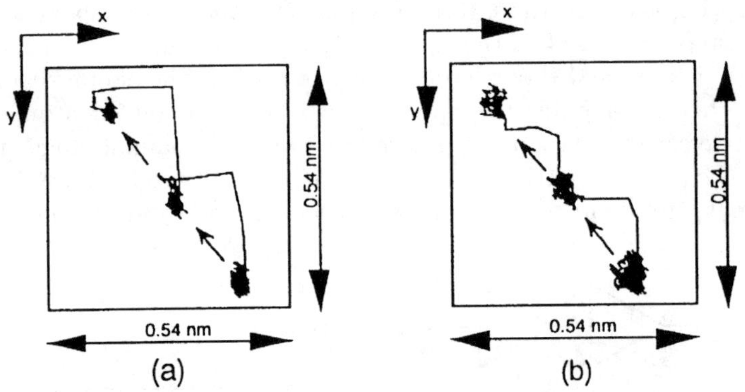

Fig. 6.9. Voltages applied to the xy scanners when the tip is positioned in registry with the crystalline lattice. (**a**) Step movement larger than the atomic spacing. (**b**) Step movement smaller than the atomic spacing

region was reported by Pohl et al. [33], and an example of differential imaging was carried out by Abraham et al. [34], and Stoll et al. [35]. Metrological application of the SPM has also been reported in the literature [36–45].

Since the technique allows positioning with lattice-regulated precision, it is a powerful tool for controlling the movement of the sample stage of a scanning probe microscope. Its merits can be fully exploited for metrological applications and imaging of weak signals in microscopy which necessitate accumulation and averaging of signals without lateral drift of the scanned area. With such an aim in mind, we developed a dual tunneling unit STM.

6.3 Fabrication

6.3.1 Fabrication of Nanometric Oscillators for Scanning Force Microscopy

A mechanical oscillator is used in non-contact scanning force microscopy (SFM) [2,46] to detect the force gradient acting at the tip, or a mass change of the tip. The detectable force resolution of a mechanical oscillator is proportional to $(KK_{\mathrm{B}}T)^{1/2}(f_0Q)^{-1/2}$ as discussed in the literature [46,47], where K is the spring constant, K_{B} is Boltzmann's constant, T is the absolute temperature, f_0 is the natural frequency of the oscillator, and Q is the quality factor. Various research groups have fabricated silicon micromachined cantilevers to suit their requirements [48–51], with the aim of either improving the force resolution or increasing the possible scanning speed. If the oscillator has a structure that allows modelling as a concentrated mass–spring model, its minimum detectable force is proportional to $(KM)^{1/2}$, where K is the spring constant and M the mass. Miniaturization of the oscillator acts favourably towards increasing the resolution, since a drastic decrease in the mass can then be achieved. The effect of miniaturization on the Q factor can be found in the literature [52,53]. Estimated values and experimental results still vary by a few orders of magnitude, possibly due to the difficulty in modelling surface effects.

We developed a novel fabrication technique in order to increase the force and mass sensitivity of the oscillator used in SFM, and if necessary, to implement a dense array of a great many oscillators. In this approach, a tip-shaped mass is supported by an elastic neck. SOI (silicon on insulator) and SIMOX (separation by implanted oxygen) wafers were used as the initial wafer. Oscillators were of the order of 1 000 nm for those made from SOI, and 100 nm for those made from SIMOX.

We aim to fabricate the tip–neck–substrate configuration as depicted in Fig. 6.10j. The various fabrication stages are depicted in Figs. 6.10a, d, g and j. To realize such a structure, we selected a silicon–silicon dioxide laminated substrate, such as SOI (silicon on insulator) and SIMOX (separation by implanted oxygen). SOI and SIMOX substrates are laminated in the following order from the bottom towards the surface: silicon, silicon dioxide, and silicon, as depicted in Figs. 6.10a, b and c. SOI substrates are usually fabricated by bonding two separate silicon wafers together and the quality of the bonding varies from one preparation method to another. In the case of SIMOX, the internal oxide layer is fabricated by oxygen implantation. Details of the fabrication process are discussed elsewhere [24,25]. By using the anisotropic etching property of silicon by KOH, tetrahedral tips can be fabricated as depicted in Figs. 6.10d, e and f. The method is an existing technique

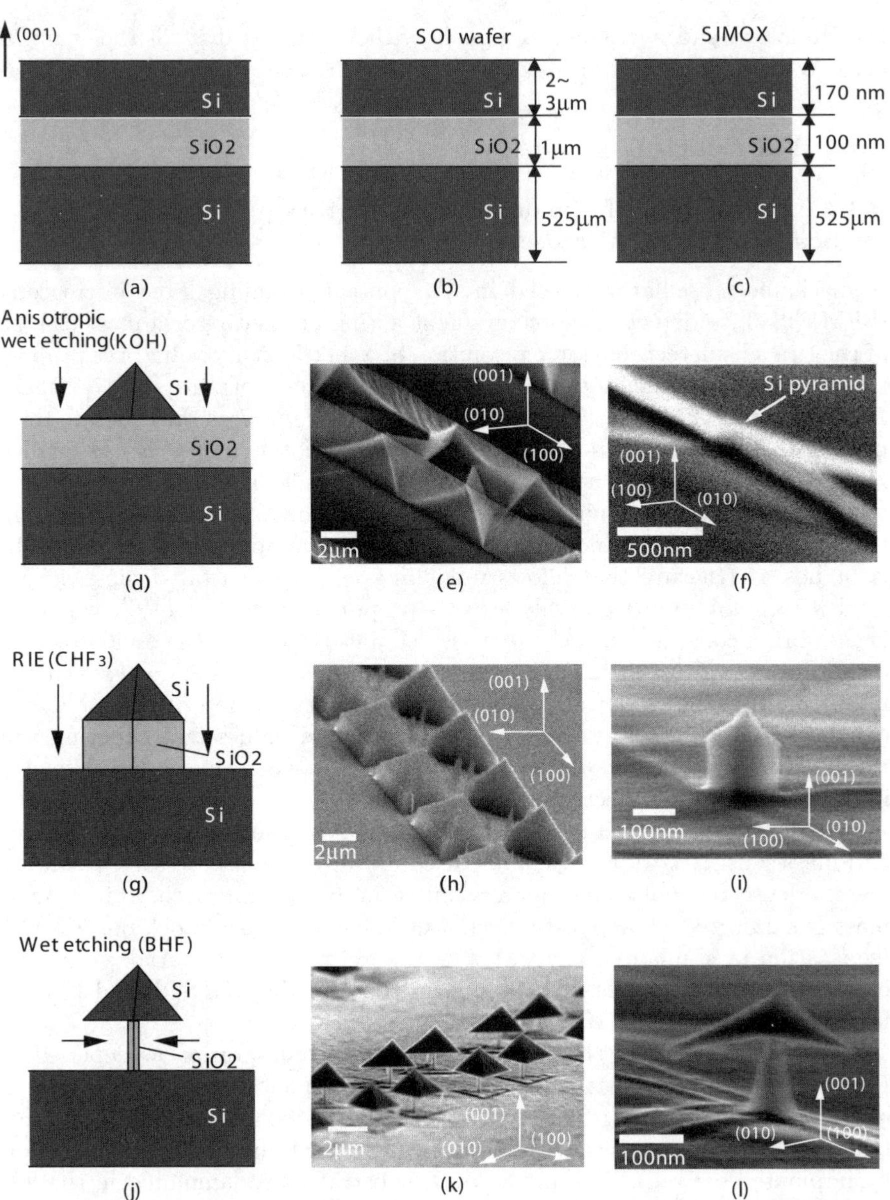

Fig. 6.10. Fabrication procedures for a nanometric tip supported by an elastic neck, where the tip is made by anisotropic etching of silicon by KOH. (**a**) Initial layered wafer, (**b**) dimensions of an SOI wafer, (**c**) dimensions of a SIMOX wafer, (**d**) tip fabrication by anisotropic etching of silicon by KOH, (**e**) image of the tip for an SOI wafer, (**f**) image of the tip for a SIMOX wafer, (**g**) reactive ion etching of columns by SF_6, (**h**) columns made on SOI, (**i**) columns made on SIMOX, (**j**) etching of the silicon dioxide layer by diluted HF, (**k**) neck formation on SOI, and (**l**) neck formation on SIMOX

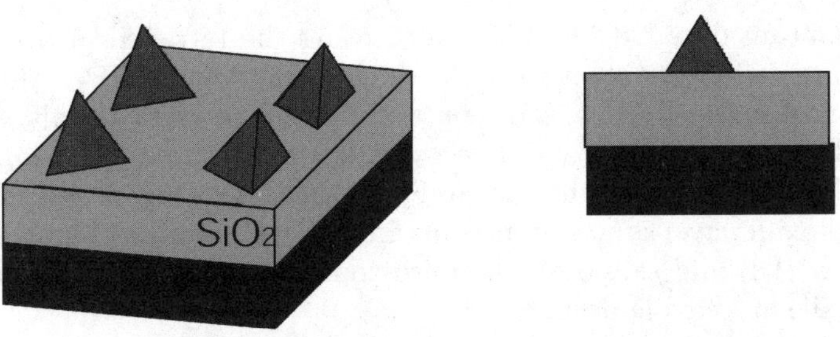

(a) Formation of Si pyramids by anisotropic etching

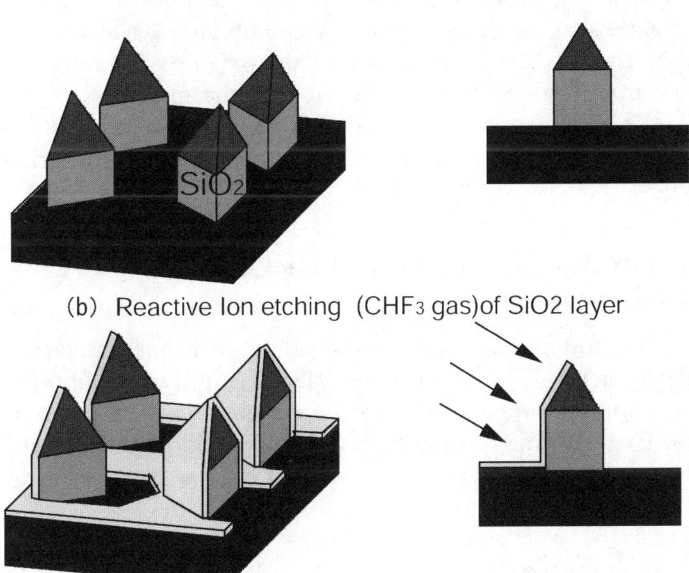

(b) Reactive Ion etching (CHF3 gas)of SiO2 layer

(c) Oblique deposition ofpoly- Si by sputtering

(d) Removal of SiO2 layer by BHF

Fig. 6.11. Fabrication process for obtaining tetrahedral tips supported by thin leaf-spring-type cantilevers. (**a**) Formation of Si pyramids by anisotropic etching. (**b**) Reactive ion etching (CHF_3 gas) of SiO_2 layer. (**c**) Oblique deposition of polysilicon by sputtering. (**d**) Removal of SiO_2 layer by BHF

for fabricating quantum dots [28,29]. After fabricating the tetrahedral tips, vertical columns are etched by reactive ion etching (RIE) with CHF_3. The tip serves as the hard mask. Vertical RIE continues until the vertical wall of the underlying silicon dioxide appears. The oxide layer measured 1 000 nm in the case of SOI, and 100 nm in the case of SIMOX (Figs. 6.10g, h and i). Then the silicon dioxide layer is wet-etched by diluted HF solution. Figures 6.10j, k and l show the final shape of the anisotropically etched silicon tip supported by the silicon dioxide neck.

Figure 6.11 depicts a method for obtaining a leaf-spring-type cantilever. This is realized by oblique deposition of polysilicon on the columns, before removal of the sacrificial silicon dioxide layer. It was possible to fabricate cantilevers with thickness as small as 20 nm. A compliant cantilever with natural frequency in the 10 MHz range may enable wearless scanning of the surface in the contact mode. Figure 6.12 shows some examples of the cantilevers fabricated by the proposed method. Cantilevers measuring 4 μm in length could be made as thin as 30 nm due to the oblique deposition technique.

6.3.2 Fabrication of Nanometric Parallel Leaf Springs for Precise Linear Motion

Figure 6.13 depicts a method for obtaining a parallel leaf spring for supporting a tip or a rectangular structure. By oblique deposition of silicon from opposing directions, polysilicon may be deposited on two opposing walls of columns supporting the tips. By removing the sacrificial silicon dioxide layer, parallel leaf springs are formed (see Fig. 6.14). This structure gives a more linear motion than cantilevers.

6.3.3 Fabrication of Millions of Cantilevers on a Centimeter Square Chip

Although the scanning probe method has very high spatial resolution, it is not time-effective because the raster scanning of the probe is mechanical. The accuracy of imaging and probe lithography will deteriorate due to drift as the time required for scanning increases. A few groups have fabricated multicantilever arrays to solve the problem and to improve efficiency [54–61]. The fabrication methods introduced in this chapter use the anisotropic etching property of silicon to obtain small structures that are well defined in both size and shape. This etching property is also well adapted for fabricating a large number of cantilevers with high uniformity and density.

Figure 6.15 depicts the fabrication process. The process resembles the method introduced above, but the etching of the SOI layer is stopped approximately 100 nm before the etching surface reaches the silicon dioxide layer. The triangular cantilever shape is also formed by anisotropic etching of silicon by KOH. This enables exact positioning of each tip on the edge

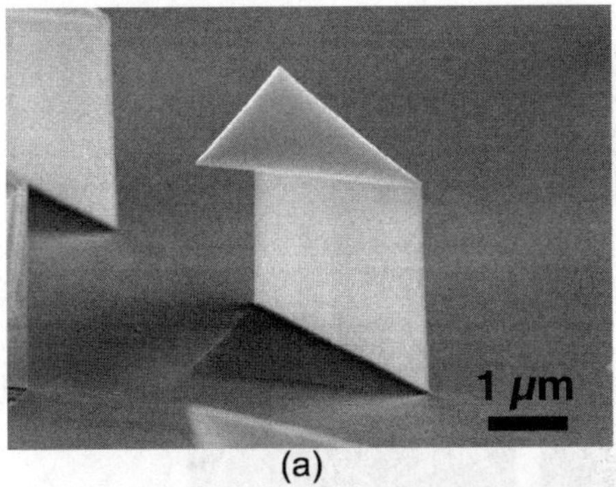

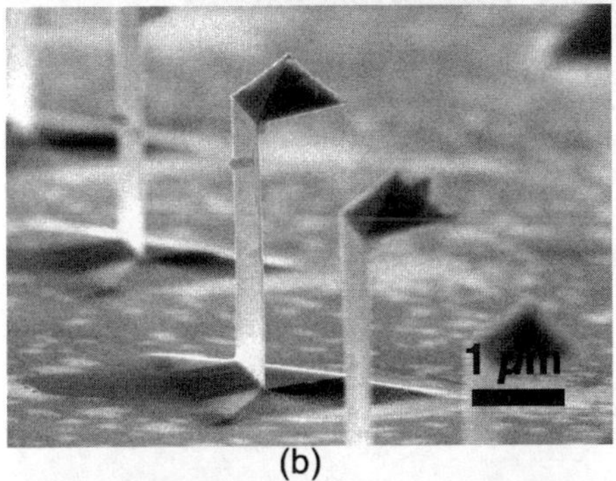

Fig. 6.12. Examples of cantilevers fabricated by oblique deposition of silicon. The size of the tetrahedral tip can be chosen by changing the SOI thickness. (**a**) Cantilever length 3 μm, thickness 80 nm, spring constant 3 N/m, natural frequency 18 MHz. (**b**) Cantilever length 3 μm, thickness 30 nm, 0.1 N/m, 3 MHz. (**c**) Cantilever length 3 μm, thickness 100 nm, 5 N/m, 22 MHz

of the triangular cantilever. The silicon dioxide layer is etched by BHF. The etching time determines the amount of overhang of the cantilevers. Figure 6.16 shows an example of a cantilever array fabricated by this method. There are approximately 1.5 million cantilevers in a centimeter square. The thickness of the cantilever is thinned by oxidation and removal by BHF until the desired value is obtained. The figure shows a case where the cantilever thickness is 80 nm.

From calculation and measurement, the spring constant of the cantilevers is 10 N/m. The calculated natural frequency is 11 MHz. Due to their reduced

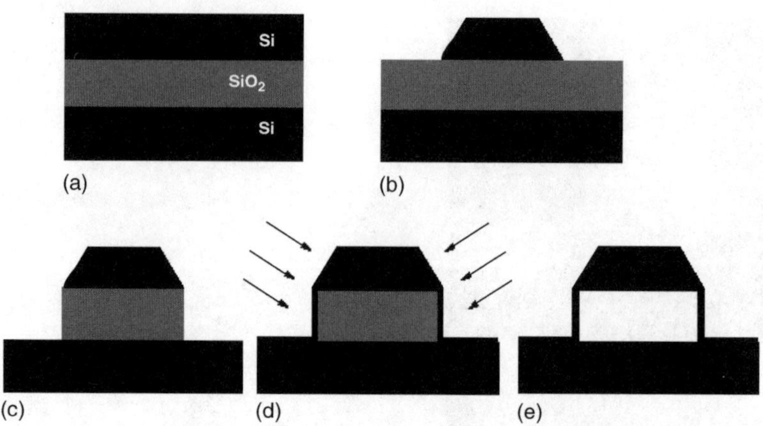

Fig. 6.13. Fabrication process for obtaining parallel leaf springs. Polysilicon is deposited on two opposing surfaces of the silicon dioxide columns. The silicon dioxide layer is etched to leave silicon tips or beams supported at both ends by a parallel leaf spring. (**a**) SO afer; (**b**) formation of Si wire; (**c**) formation of Si-SiO_2 column; (**d**) oblique deposition of poly-Si by sputtering; (**e**) removal of SiO_2 layer of BHF

Fig. 6.14. Example of the parallel leaf spring structure

thickness, cantilevers with high natural frequency and reasonable compliance were obtained despite their size. Cantilevers with spring constants adapted for non-contact or contact mode scanning force microscopy can be obtained by choosing the cantilever length and thickness around the values shown in Fig. 6.16b.

6.3.4 Strength Measurement of the Nano-Oscillator

Using an SFM operating inside a scanning electron microscope, as depicted in Fig. 6.17, a force was applied to the nanometric oscillator to measure the strength of the neck [30]. The tip pointing downwards is the tip of a conventional SFM cantilever. In some cases, hooks were made on the SFM cantilever tip by electron beam irradiation to anchor the tip against the nano-oscillator.

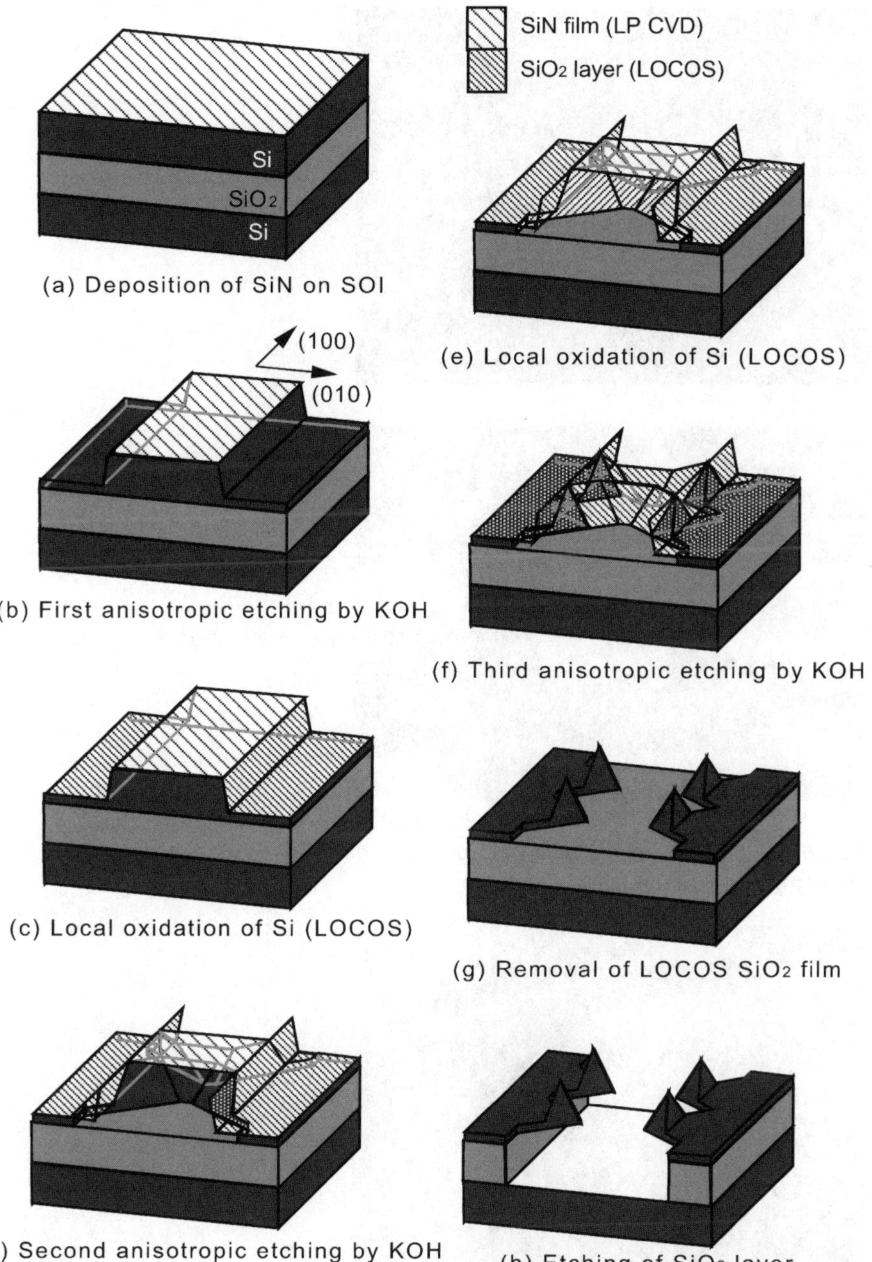

Fig. 6.15. Process for fabricating millions of cantilevers on a centimeter square chip

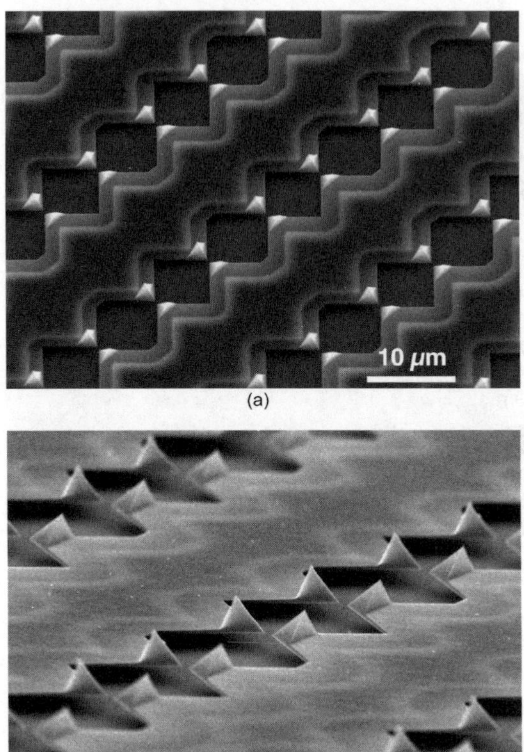

Fig. 6.16. Example of millions of cantilevers per centimeter square, fabricated by anisotropic etching of silicon by KOH

Fig. 6.17. Scanning force microscope operating in a commercial scanning electron microscope. The tip of a conventional AFM cantilever is seen pointing downwards

The maximum shear force applicable to the oscillator before rupture at the neck was 5×10^{-6} N and the maximum normal force was $2\text{–}3 \times 10^{-6}$ N when applied to one of the lower edges of the tip. The neck area at the point of failure measured approximately 10^{-14} m^2. Considering the effect of leverage when the force was applied, the strength of the neck was not smaller than 1 GPa for both shear and normal forces. For a neck measuring 10 nm in diameter, its strength would be of the order of 100 nN in both the shear and normal directions, implying that reasonable strengths can still be obtained for even smaller oscillators.

6.4 Characterization

We have built a scanning probe microscope that mounts on a commercial scanning electron microscope base [30]. The scanning electron microscope is used to position the scanning probe against the sample. By such a combination, probing and force application to the sample can be executed at the desired location and in the intended direction. Figure 6.18 shows a nanometric silicon dioxide neck being bent by a commercial scanning force microscope tip. Figure 6.19 is a metal tip pressing one of the millions of cantilevers. From the deflection of the scanning force microscope cantilever, the strength of the sample can be evaluated. For example, a silicon–silicon dioxide interface with a surface area of 10^{-14} m^2 had a strength of 1 GPa. Ultrasonic probes and scanning near-field optical microscope probes can also be positioned with the apparatus.

Fig. 6.18. Nanocantilever under strength measurement by SEM AFM. The tip and base are made of silicon and the neck of silicon dioxide. The silicon–silicon dioxide interface has a strength of at least 1 GPa

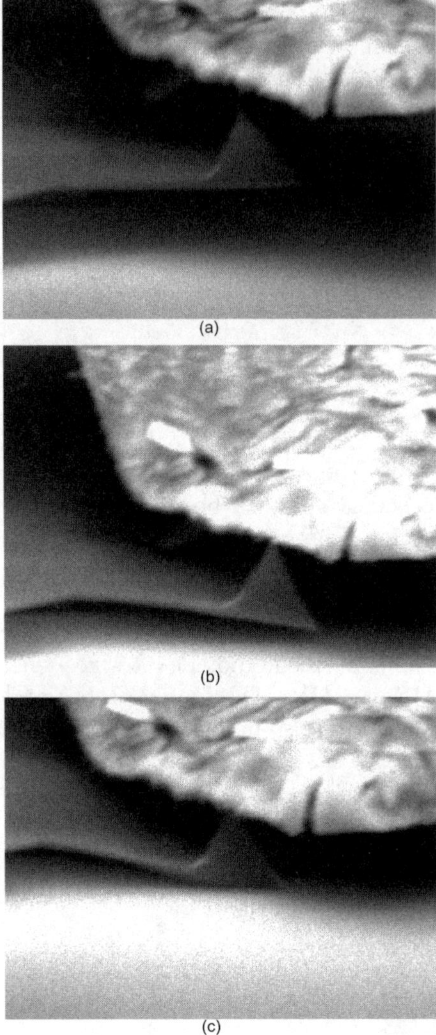

Fig. 6.19. One of the millions of silicon cantilevers is put under strength measurement by an STM tip. The cantilever deforms elastically. Its spring constant was 10 N/m when measured by an AFM. Calculated values are: spring constant 10 N/m, natural frequency 11 MHz

6.5 Conclusion

This chapter gave an overview of some topics in the field of scanning probe methods and nano- to microdetectors. Some of the challenging issues in these topics are:

- widening the frequency range of observation and spectroscopic studies of various quantities,
- controlling and acquiring data from millions of detectors,
- implementing novel imaging techniques with new detectors.

Acknowledgments. The paper is a collection of various projects the author was lucky enough to carry out with other researchers and students. The author would like to thank Y. Hoshi, T. Saito, T. Kawagishi, A. Kato (tribology), T. Higuchi, H. Kougami (control), D. Saya, H. Fujita, H. Toshiyoshi, G. Hashiguchi, H.-J. Hug, H.-J. Guntherodt (nanometric oscillators), and K. Fukushima (SEM AFM).

References

1. G. Binnig, H. Rohrer: *Scanning Tunneling Microscopy*, Helv. Phys. Acta **55**, 726 (1982)
2. G. Binnig, C.F. Quate, C. Gerber: *Atomic Force Microscope*, Phys. Rev. Lett. **12**, 930 (1986)
3. D.W. Pohl: IBM J. Res. Develop. **30**, 417 (1986)
4. D.W. Pohl, D. Courjon (Eds.): *Near Field Optics*, NATO ASI Series E, 242 (Kluwer Academic Publishers, Dordrecht 1993)
5. P. Gutter, H.J. Mamin, D. Rugar: in *Scanning Tunneling Microscopy II*, ed. by R. Wiesendanger and H.-J. Guntherodt (Springer, Berlin 1992) p. 151
6. D. Rugar, H.J. Mamin, P. Guethner, S.E. Lambert, J.E. Stern, I.R. McFadyen, T. Yogi: J. Appl. Phys. **68**, 1169 (1990)
7. C. Shonenberger, S.F. Alvarado: Z. Phys. B **80**, 373 (1990)
8. R. Kaneko, K. Nonaka, K. Yasuda: *Scanning Tunneling Microscopy and Atomic Force Microscopy for Microtribology*, J. Vac. Sci. Technol. **A6-2**, 291 (1988)
9. E. Meyer, R. Overney, D. Brodbeck, L. Howald, R. Luthi, J. Frommer, H.-J. Guntherodt: *Friction and Wear of Langmuir–Blodgett Films Observed by Friction Force Microscopy*, Phys. Rev. Lett. **69**, 1777 (1992)
10. C. Mate, G. McClelland, R. Erlandsson, S. Chang: *Atomic Scale Friction of a Tungsten Tip on a Graphite Surface*, Phys. Rev. Lett. **59**, 1942 (1987)
11. G. Meyer, N.M. Amer: *Simultaneous Measurement of Lateral and Normal Forces with an Optical-Beam-Deflection Atomic Force Microscope*, Appl. Phys. Lett. **57**, 2089 (1990)
12. R.C. Barrett, C.F. Quate: Ultramicroscopy **42–44**, 262 (1992)
13. S. Watanabe, K. Hane, T. Ohye, M. Ito, T. Goto: *Electrostatic Force Microscope Imaging Analyzed by the Surface Charge Method*, J. Vac. Sci. Technol. B **11**, 1774 (1993)
14. C. Shafai, D.J. Thomson, M. Simard-Norman: J. Vac. Sci. Technol. B **12**, 378 (1994)
15. H. Kawakatsu, T. Saito: *Scanning Force Microscopy with Two Optical Levers for Detection of Deformations of the Cantilever*, J. Vac. Sci. Technol. B **14**, 872 (1996)
16. H. Kawakatsu, H. Bleuler, T. Saito, H. Kougami: *Dual Optical levers for Atomic Force Microscopy*, Jpn. J. Appl. Phys. **34**, 3400 (1995)
17. H. Kawakatsu, T. Saito, H. Kougami, P. Blanalt, M. Kawai, M. Watanabe, N. Nishioki: *Detecting and Controlling Forces in Atomic Force Microscopy with Multidegrees of Freedom*, J. Vac. Sci. Technol. B **12**, 1686 (1994)
18. H. Kawakatsu, T. Higuchi: J. Vac. Sci. Technol. A **8**, 319 (1990)

19. H. Kawakatsu, Y. Hoshi, T. Higuchi, H. Kitano: J. Vac. Sci. Technol. B **9**, 651 (1991)
20. H. Kawakatsu, Y. Hoshi, H. Bleuler, H. Kougami, M. Bossardt, N. Vezzin: Appl. Phys. A **66**, S853 (1998)
21. H. Kawakatsu, H. Kougami: *Automated Calibration of the Sample Image Using Crystalline Lattice for Scale Reference in Scanning Tunneling Microscopy*, J. Vac. Sci. Technol. B **14**, 11 (1996)
22. T. Fujii, M. Suzuki, T. Higuchi, H. Kougami, H. Kawakatsu: J. Vac. Sci. Technol. B **13**, 1112 (1995)
23. D. Saya, H. Toshiyoshi, H. Fujita, G. Hashiguchi, H. Kawakatsu: Jpn J. Appl. Phys. **39** 6B, 3793 (2000)
24. H. Kawakatsu, H. Toshiyoshi, D. Saya, H. Fujita: Jpn J. Appl. Phys. **38**, 3962 (1999)
25. H. Kawakatsu, H. Toshiyoshi, D. Saya, K. Fukushima, H. Fujita: J. Vac. Sci. Technol. **18**, 607 (2000)
26. H. Kawakatsu, H. Toshiyoshi, D. Saya, K. Fukushima, H. Fujita: J. Vac. Sci. Technol (submitted STM99)
27. H. Kawakatsu, H. Toshiyoshi, D. Saya, K. Fukushima, H. Fujita: Appl. Surf. Sci. **157**, 320 (2000)
28. G. Hashiguchi, H. Mimura: Jpn. J. Appl. Phys. **33**, L1649 (1994)
29. G. Hashiguchi, H. Mimura: Jpn. J. Appl. Phys. **34**, 1493 (1995)
30. K. Fukushima, D. Saya, H. Kawakatsu: Jpn. J. Appl. Phys. **39** 6B, 3747 (2000)
31. H. Kawakatsu, Y. Hoshi, T. Kawagishi, A. Kato: submitted
32. Y. Hoshi, H. Kawakatsu: Jpn. J. Appl. Phys. **39** 6B (2000)
33. D.W. Pohl, R. Moller: Rev. Sci. Instrum. **59**, 840 (1988)
34. D.W. Abraham, C.C. Williams, H.K. Wickramasinghe: J. Microsc. **152**, 599 (1988)
35. E.P. Stoll, J.K. Gimzewski: *Fundamental and Practical Aspects of Differential Scanning Microscopy*, J. Vac. Sci. Technol. B **9**, 643 (1991)
36. M. Aketagawa, K. Takada, S. Sasaki, K. Kobayashi, S. Suzuki, K. Yamada, Y. Nakayama: in *Progress in Precision Engineering and Nanotechnology*, ed. by H. Kunzmann et al. (PTB-Verlag, Braunschweig 1997) p. 49
37. A. Torii, M. Sasaki, K. Hane, S. Okuma: *The Feasibility of a Precise Linear Displacement Encoder Using Multiple Probe Force Microscope*, Int. J. of Japan Soc. for Precision Engineering **27**, 367 (1993)
38. H. Shang, T. Higuchi, N. Nishioki: *Dual Tunneling-Unit Scanning Tunneling Microscope for Length Measurement Based on Crystalline Lattice*, J. Vac. Sci. Technol. B **15** (1), 174–177 (1997)
39. H. Zhang, F. Huang, T. Higuchi: *Dual Unit Scanning Tunneling Microscope–Atomic Force Microscope for Length Measurement Based on Reference Scales*, J. Vac. Sci. Technol. B **15**, 780 (1997)
40. C. Teague: J. Vac. Sci. Technol. B **7**, 1898 (1989)
41. T.V. Vorburger, J.A. Dagata, G. Wilkening, K. Iizuka: Annals of the CIRP **46**, 597 (1997)
42. S. Sakuta, K. Youcerf-Toumi: Precision Engineering **62**, 403 (1996)
43. J.E. Griffith, D.A. Grigg: *Dimensional Metrology with Scanning Probe Microscopes*, J. Appl. Phys. **74**, R83 (1993)
44. M. Stedman: *The Performance and Limits of Scanning Probe Microscopes*, Inst. Phys. Conf. Ser. No. 130, Chap. 4; M. Stedman: J. Microscopy **152**, 611 (1988)

45. J.S. Villarrubia: Surf. Sci. **321**, 287 (1994); J. Res. Natl. Inst. Stand. & Technol. **102** (4), 425 (1997)
46. T.R. Albrecht, P. Gutter, D. Horne, D. Rugar: J. Appl. Phys. **69**, 668 (1991)
47. T.D. Stowe, K. Yasumura, T.W. Kenny, D. Botkin, K. Wago, D. Rugar: Appl. Phys. Lett. **71**, 288 (1997)
48. D.A. Walters, J.P. Cleveland, N.H. Thomson, P.K. Hansma, M.A. Wendman, G. Gurley, V. Elings: Rev. Sci. Instrum. **67**, 3583 (1996)
49. K. Wago, O. Zuger, J. Wegener, R. Kendrick, C.S. Yannoni, D. Rugar: Rev. Sci. Instrum. **68**, 1823 (1997)
50. B.W. Chui, T.D. Stowe, T.W. Kenny, H.J. Mamin, B.D. Terris, D. Rugar: Appl. Phys. Lett. **69**, 2767 (1996)
51. G.T. Paloczi, B.L. Smith, P.K. Hansma, D.A. Walters, M.A. Wendman: Appl. Phys. Lett. **73**, 1658 (1998)
52. J. Yang, T. Ono, M. Esashi: in *Proc. of the 13th Annual International Conference on MEMS*, 00CH36308, Miyazaki, Japan (2000) p. 235
53. Vu. Thien Binh, N. Garcia, A.L. Levaanuyk: Surf. Sci. **301**, L224 (1994)
54. S.A. Miller,
55. M. Lutwyche, C. Andreoli, G. Binnig, J. Brugger, U. Drechsler, W. Haeberle, H. Rohrer, H. Rothuizen, P. Vettiger, G. Yaralioglu, C.F. Quate: *5×5 2D AFM Cantilever Arrays: A First Step Towards a Terabit Storage Device*, Sens. & Actuat. A **73**, 89 (1999)
56. P. Vettiger, J. Brugger, M. Despont, U. Drechsler, U. Durig, W. Haeberle, M. Lutwyche, H. Rothuizen, R. Stutz, R. Widmer, G. Binnig: *Ultrahigh Density, High-Data-Rate NEMS-Based AFM Data Storage System*, Micro. Eng. **46**, 11 (1999)
57. E.M. Chow, H.T. Soh, H.C. Lee, J.D. Adams, S.C. Minne, G. Yaralioglu, A. Atalar, C.F. Quate, T.W. Kenny: *Integration of Through-Wafer Interconnects with a Two-Dimensional Cantilever Array*, Sens. & Actuat. A **83**, 118 (2000)
58. P.-F. Indermuhle, G. Schurmann, G.-A. Racine, N.F. de Rooij: *Fabrication and Characterization of Cantilevers with Integrated Sharp Tips and Piezoelectric Elements for Actuation and Detection for Parallel AFM Applications*, Sens. & Actuat. A **60**, 186 (1997)
59. M. Despont, J. Brugger, U. Drechsler, U. Duerig, W. Haeberle, M. Lutwyche, H. Rothuizen, R. Stutz, R. Widmer, G. Binnig, H. Rohrer, P. Vettiger: *VLSI-NEMS Chip for Parallel AFM Data Storage*, Sens. & Actuat. A **80**, 100 (2000)
60. H.P. Lang, M.K. Baller, R. Berger, Ch. Gerber, J.K. Gimzewski, F.M. Battiston, P. Fornaro, J.P. Ramseyer, E. Meyer, H.-J. Guntherodt: *An Artificial Nose Based on a Micromechanical Cantilever Array*, Analytica Chimica Acta **393**, 59 (1999)
61. H.P. Lang, R. Berger, C. Andreoli, J. Brugger, M. Despont, P. Vettiger, Ch. Gerber, J.K. Gimzewski, J.P. Ramseyer, E. Meyer, H.-J Guntherodt: *Sequential Position Readout from Arrays of Micromechanical Cantilever Sensors*, Appl. Phys. Lett. **72**, 383 (1998)

7 Micromachined Scanning Tunneling Microscopes and Nanoprobes

Hiroyuki Fujita, Yasuo Wada, Dai Kobayashi, and Gen Hashiguchi

The magnitude of a tunneling current that passes across a vacuum gap changes by about one decade as the gap changes by 0.1 nm. The very high sensitivity of electron tunneling with respect to the gap distance can be used to sense very small changes in the gap distance. An STM (scanning tunneling microscope) [1] is a distinctive application of this very high sensitivity of electron tunneling. The STM has been utilized as a powerful and convenient tool for observing and manipulating individual atoms and molecules on the nanoscopic scale. However, a conventional STM has dimensions over 10^8 times larger than the positioning accuracy required of it, and this means that STMs encounter the following problems:

- they are liable to be affected by vibrations due to the heavy weight and large size of the actuator, so that very good anti-vibration equipment is needed;
- they are relatively expensive due to the cost of components and the need for precise assembly and anti-vibration equipment;
- it is difficult to reduce their size due to the small deformation limit ($<$ 0.01%) of the piezoelectric material, and a minimum size of several mm is necessary;
- they are subject to instability and nonlinearity due to creep and hysteresis in the piezoelectric actuator.

Micromachining can realize miniature tunneling devices with very small dimensions and mass, resulting in the following advantages:

- high robustness to vibrations from the environment and small thermal drift, eliminating the need for anti-vibration equipment;
- low cost, because micromachine fabrication relies on the ULSI (ultralarge scale integrated circuit) microfabrication technology, and this makes it possible to fabricate a large number of chips at the same time;
- integration capability, in which both mechanical and electronic parts can be fabricated on the same chip with very small dimensions;
- use under special conditions such as high or low temperatures, or very restricted space;
- applicability in microscopy and built-in micro-displacement sensors.

Micromachined tunneling units [2] and STMs [3,4] have been previously reported. Some of them require assembly [2] and coarse adjustment of the opposing surface [3,4]. In all cases, the coarse positioner was not made by micromachining. In order to take advantage of miniaturization, the total size of the system must be minimized. Moreover, no assembly should be involved, because it restricts miniaturization. Miniaturized design makes assembly difficult. The tunneling unit must therefore be designed in such a way that thorough miniaturization is possible and no assembly is required.

Silicon-based surface micromachining is the most suitable way of miniaturizing a tunneling unit. This technology makes it possible to integrate a microactuator, a tip and an opposing wall, which corresponds to the STM sample, in the same plane in the same process. By integrating other microstructures such as the proof mass of an accelerometer or an AFM cantilever instead of a fixed opposing wall, various sensors can be obtained. Single substrate design saves the area of the fringe for wafer bonding. Since the tunneling tips of adjacent units can be put close to each other, this design will constitute one way of realizing a multiprobe scanning microscope.

The major research laboratories that have led research towards monolithically integrated STMs are the Institute for Industrial Science at the University of Tokyo [5–8] and the Advanced Research Laboratory, Hitachi Ltd. [9–11]. This chapter mainly describes advances in the design, fabrication and characterization of micromachined STM (micro-STM) technology with reference to the results reported from these institutions. It also deals with the future application of micro-STMs in advanced lithography equipment for ULSI fabrication and ultrahigh-density data storage devices. These applications would fully utilize the advantages of the micro-STM, because it would make massively parallel operation possible and fulfill the requirements of both sub-100 nm lithography technology and tera-bit data storage. Finally, we describe advanced applications of micro-STM to the understanding of physical and chemical phenomena occurring between the tip and sample of the STM under an extremely high electric field.

7.1 Operating Principles and Basic Structure

A micromachined STM needs actuators to move the tip and control its position, and actuators based on several principles have been proposed so far for generating the necessary force:

- electrostatic actuators,
- piezoelectric actuators,
- thermal actuators.

The advantages and disadvantages of these actuation methods are compared in Table 7.1. This clearly brings out the overall advantages of the electrostatic actuator, such as compatibility with the ULSI fabrication process, small size and fast response. Therefore, almost all micro-STM studies have assumed the use of electrostatic actuators [5–11]. One reason for this choice is that this

Table 7.1. Advantages and disadvantages of micro-STM actuators. E: excellent, G: good, F: fair, P: poor

Type	Displacement	Force	Stability	Size	Response	ULSI compatibility
Electrostatic	E	G	G	E	E	E
Piezoelectric	F	E	P	G	G	F
Thermal	F	E	E	G	P	G

kind of actuator has a stroke of motion large enough to cover the initial tip–wall gap. The initial gap ranges from 0.5 to several micrometers, depending on whether the tip is sharpened or not. The other reason is that a comb drive has no hysteresis and not very pronounced nonlinearity. The displacement of a comb drive is proportional to the square of the driving voltage and can be cancelled by using a pair of comb drives with identical characteristics.

The operating principle of an electrostatically actuated micro-STM is explained in the schematic drawing of Fig. 7.1a. A tunneling tip, an opposing wall and a microactuator are integrated on a single chip. Figure 7.1b is an SEM picture of the developed tunneling unit. The tip and actuator are released from the substrate while the driving pad and the wall are fixed to it. The movement of the main body is generated by electrostatic comb actuators by applying voltages between the two comb electrodes. The main body is suspended by four springs which hold the main body tightly, guaranteeing a strictly constrained one-dimensional motion of the main body. The voltage across the comb actuator is 10–50 V. This is low enough to operate thin film devices because the breakdown voltages of insulators used in ULSIs, such as silicon dioxide and silicon nitride, are of the order of 1 V/nm, and an insulator layer 50–100 nm thick will therefore prevent electrical shorts.

7.2 Micro-STM Design Considerations

7.2.1 Basic Design of Electrostatic Actuators

The force (F) generated by the electrostatic actuator is given by [5,6]

$$F = \frac{1}{4}\frac{L\epsilon_0 t}{s_2(s_1+s_2)}V^2 , \tag{7.1}$$

where s_1, s_2, L and t are the width, separation, total length and thickness of the comb actuator, respectively, as shown schematically in Fig. 7.2, whilst ϵ_0 is the permittivity and V the voltage between the two comb electrodes. If a balanced actuator mode is employed [9], (7.1) is expressed by

$$F = \frac{1}{4}\frac{L\epsilon_0 t}{s_2(s_1+s_2)}V_{\mathrm{drive}}V , \tag{7.2}$$

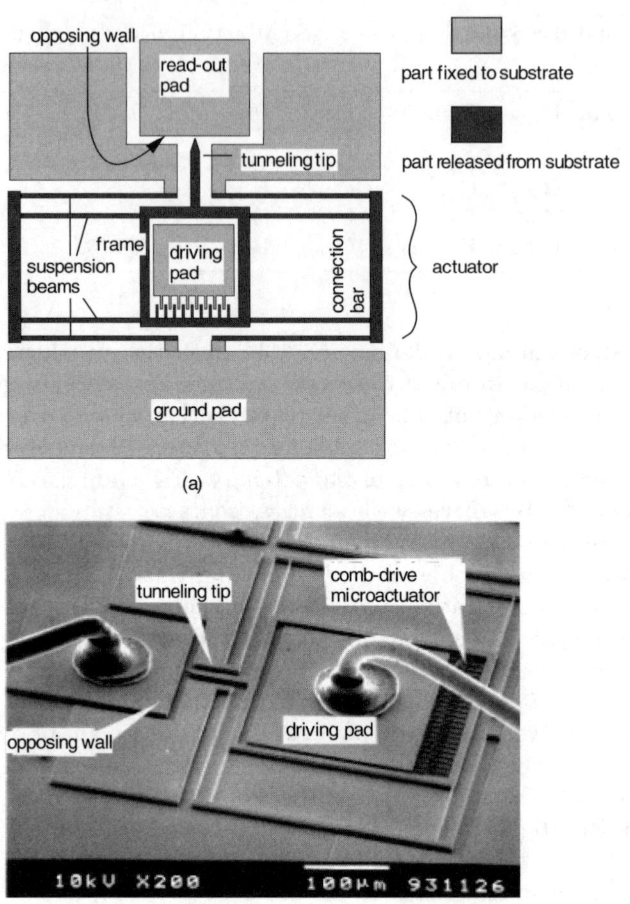

Fig. 7.1. (**a**) Schematic drawing and (**b**) SEM picture of the micro-STM

where V_{drive} is the voltage applied on the middle electrode. Equations (7.1) and (7.2) indicate clearly that the force generated by the electrostatic comb actuator is inversely proportional to the square of the minimum dimension, l_{min}, when $l_{\text{min}} = s_1 = s_2$. Therefore, it should only be possible to make a high performance actuator by using the smallest possible dimensions. These equations also indicate that the confronting electrode area (Lt) is reduced by a factor of n^2, if the size is reduced by a factor of n. In other words, the scaling principle of the electrostatic comb actuator is as follows: if the size is reduced by a factor of n, then the force is kept constant if the voltage is kept constant. It is also suggested that the confronting electrode area (Lt) has to be large enough to generate a large force to cope with the electrostatic force of about 10^{-6} N between the tip and the sample [8,9].

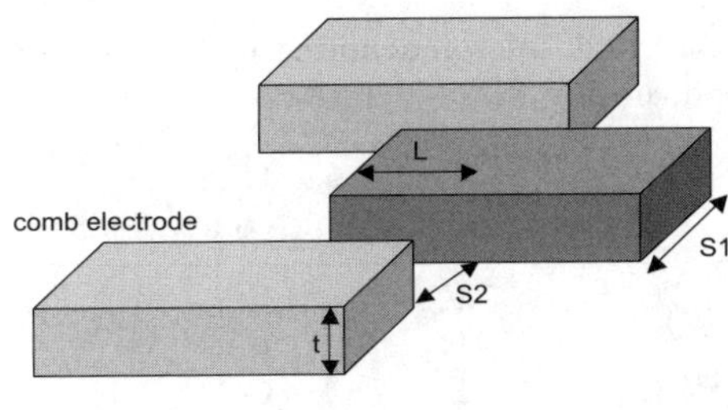

Fig. 7.2. Schematic illustration of the electrostatic comb actuator

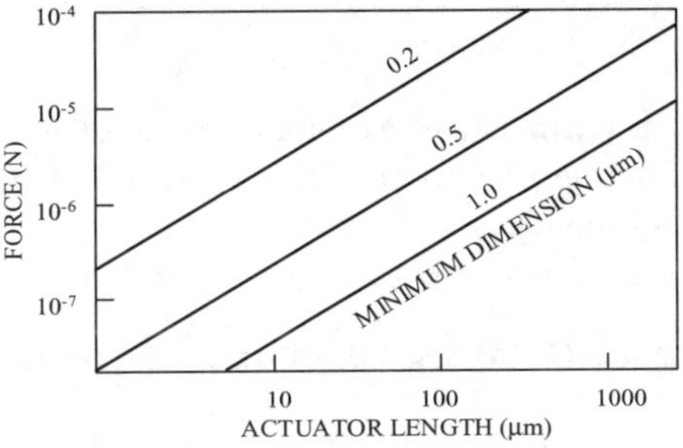

Fig. 7.3. Dependence of force F generated in the electrostatic actuator on the minimum dimension between the electrodes, as estimated from (7.1) and (7.2)

The force generated in the electrostatic actuator is estimated using (7.1) and (7.2), and the dependence of the force on the minimum dimension $l_{\min}$ between the comb electrodes was calculated as shown in Fig. 7.3 [9]. The results indicate that the minimum dimension between the comb electrodes should lie in the sub-0.5 μm range in order to make the total dimension reasonably small, since the total size of the actuator depends almost linearly on the square of the minimum dimension of the actuator, if it is to generate the same force F.

7.2.2 Vibration Frequency of the Micro-STM

The intrinsic vibration frequency f_{res} of a micro-STM is expressed by [6]

$$f_{\mathrm{res}} = \frac{\sqrt{k/m}}{2\pi} , \tag{7.3}$$

where k and m are the spring constant of the beam and the mass of the vibrating structure, respectively. Although the actual structure is more complicated than the model, (7.3) can be applied to the first order of approximation [6]. The resonant frequency is of the order of 10–100 kHz. Therefore, a micro-STM does not usually need external vibration isolators, if it is small enough and the design is adequate.

A further estimate of the effect of vibrations on the stability of a micro-STM has also been discussed in the following way [5]. When an acceleration of 1 m/s^2 is applied to a micro-STM, an inertial force of the order of 1×10^{-10} N is induced on the main body of the micro-STM. This results in a displacement of the tip position by about 0.5 nm. In addition, a typical high frequency

Table 7.2. Comparison between surface and bulk micromachining technology for micro-STM actuators. E: excellent, G: good, F: fair

Type	Force	Size	Shape flexibility	Response	ULSI compatibility
Surface micromachine	G	E	E	E	E
Bulk micromachine (wet)	E	F	F	E	F
Bulk micromachine (dry)	E	G	E	E	G

vibration and amplitude of 100 Hz and 1 µm cause a displacement of less than 0.2 nm [5], which should be small enough to operate the micro-STM in an ambient without anti-vibrational equipment.

7.3 Surface Micromachining and Bulk Micromachining

So far two technologies, surface micromachining technology and bulk micromachining technology, have been proposed to produce electrostatically actuated micro-STMs. Table 7.2 compares these two technologies. The advantage of surface micromachining technology is its compatibility with the present ULSI fabrication technologies, especially patterning, doping, film formation and metallizing technologies. In other words, surface micromachining technology can use state-of-the-art ULSI technology, and the most advanced technology can be integrated on the chip. Therefore, surface micromachining technology enables a very flexible fabrication, in which all peripheral circuits can also be integrated on the same chip as the micromachined part. Another type of surface micromachining is based on an SOI (silicon-on-insulator) substrate. A thin film of single-crystal silicon is attached to an SiO_2 layer on top of the substrate. Therefore, microstructures are made of single-crystal silicon in this case. Both dry etching and crystallographically anisotropic wet etching (etchants KOH, TMAH, etc.) are applicable. In the latter etching method, microstructures are defined very precisely by the silicon crystal planes, with approximately 10 nm accuracy.

The advantages of bulk micromachining technology are that it can be used to build very stiff structures, essential for a rigid micromachine, and that it is possible to generate a very strong force from the micromachine because of the large available electrode area. The recent development of deep RIE (reactive ion etching) technology, which is capable of fast, vertical and deep etching of silicon, has enabled us to fabricate high-aspect-ratio microstructures. (The aspect ratio of a structure is the ratio of its height to its width.) The maximum aspect ratio is 20–30 and the height of structures is 200–500 µm. This technology is suitable for strong and rigid microactuators with typical sizes of 100–1 000 µm.

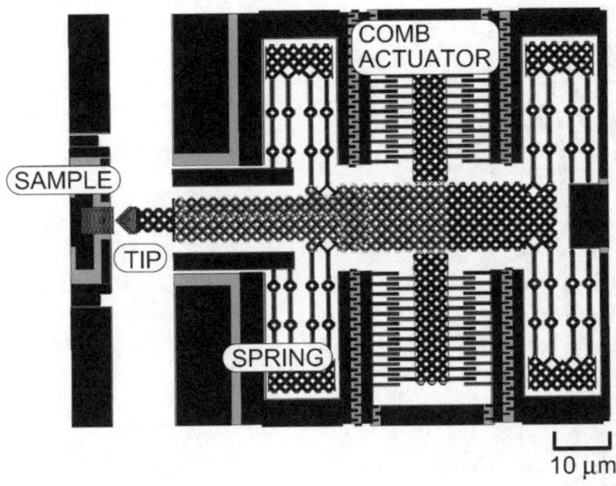

Fig. 7.4. Schematic drawing of the electrostatically actuated 1D micro-STM

Thin and sharp probes are need to investigate nanoscopic phenomena. In addition, structural rigidity and generated force must be high enough to overcome external forces, such as meniscus force and atomic force. Therefore, surface micromachining was used to make nanoprobes, while bulk micromachining was applied to microstructures and microactuators. The way to combine these processes will be described in detail later.

7.4 Micro-STM Fabrication Technology

7.4.1 Surface Micromachined STM Chip Fabrication Process

This section details the fabrication process for a micro-STM using surface micromachine technology, following the schematic design for a one-dimensional (1D) micro-STM shown in Fig. 7.4 [9]. Micro-STM chips were fabricated using ULSI fabrication technology with a minimum dimension of 0.4 μm and an alignment accuracy of 0.1 μm. Three mask layers were employed in the fabrication: a poly-Si etching layer (POLY), a through-hole layer (TH) and a metal layer (ME). Figure 7.5 shows a scanning electron micrograph (SEM micrograph) of the fabricated 1D micro-STM, which has a total size of about 200 μm square.

The details of the fabrication process and the corresponding schematic cross-section of the micro-STM are as follows. A 1 μm thick polycrystalline silicon (poly-Si) layer was deposited on a 100 nm thick silicon nitride layer (Si_3N_4) formed on an Si(100) oriented substrate, and patterned with a POLY layer, as shown in Fig. 7.6a. Next, a further 100 nm thick Si_3N_4 layer was deposited on the structure and etched using a TH mask, as indicated in Fig. 7.6b. A 500 nm thick gold layer was then deposited on the structure by means of an ME mask. This forms the tip–sample region and the wiring patterns, as depicted in Fig. 7.6c. Finally, the structures, including the poly-

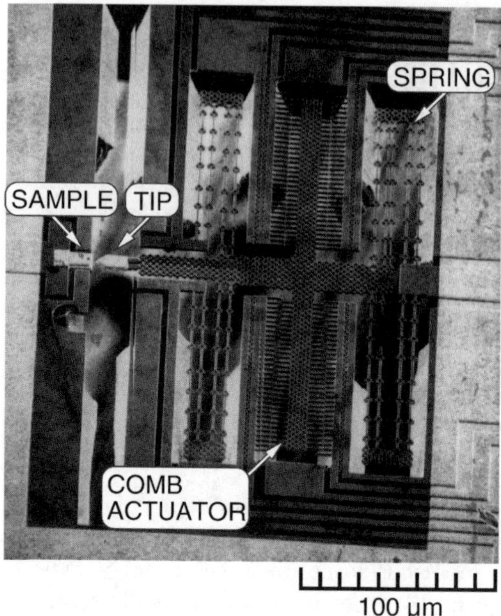

Fig. 7.5. Scanning electron micrograph of the 1D micro-STM. Total size approximately 200 μm square

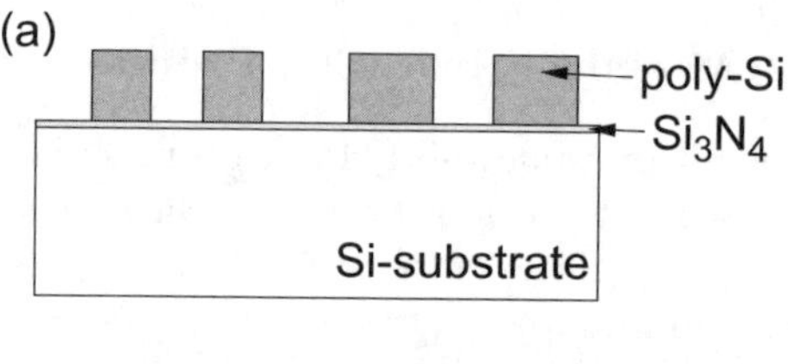

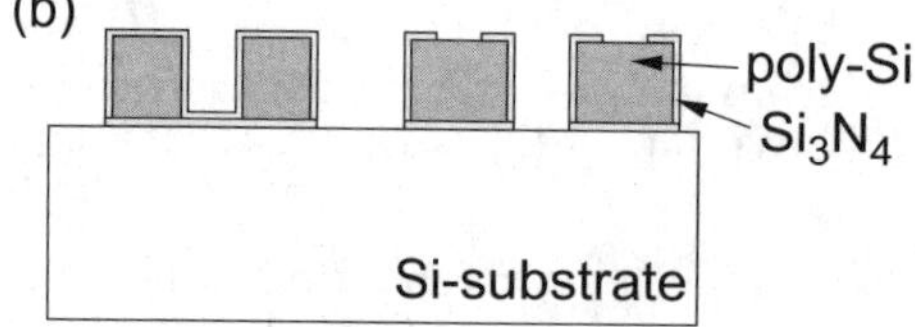

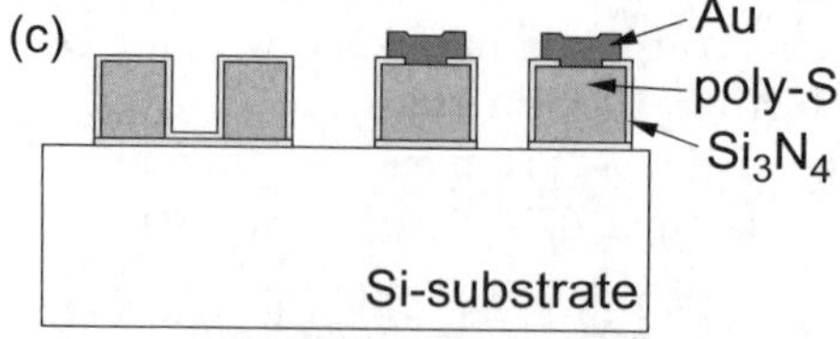

Fig. 7.6. Schematic cross-section of the micro-STM fabrication process. (**a**) Poly-Si etching layer (POLY), (**b**) through-hole layer (TH), (**c**) metal layer (ME)

Si main body and the gold tip–sample regions, were separated from the Si substrate by potassium hydroxide (KOH) etching of the substrate.

The micro-STM is made up of four main parts:

- comb actuators,
- springs,
- main body,
- tip and sample,

as shown in the SEM micrograph in Fig. 7.5. The main body, the comb actuators and the springs were made of a 1 μm thick poly-Si layer covered with a 100 nm thick Si_3N_4 layer, while the tip and the sample regions were made of a 500 nm thick gold (Au) layer. The Si_3N_4 layer protects the poly-Si layer from being etched by the KOH solution which is used to etch the Si substrate. The comb actuators were fabricated using the smallest possible line and space width available, in order to generate the highest possible force, according to (7.1) and (7.2). The micro-STMs were electrically connected to the STM controller via eight bonding pads on the chip. Under tunneling conditions, the bias voltage and current between the tip and sample measure around 1 V and 1 nA, respectively, corresponding to a tunneling resistance of about 1 G ohm. Therefore, such parameters as the parasitic resistance and controller circuit impedance have to be designed appropriately for high performance operation. In addition, two- and three-dimensional movements of the micro-STM were made possible by installing corresponding electrostatic actuators [12].

The micro-STM structure can accommodate much thicker films for higher performances by applying suitable processing technologies [5]. These include isotropic reactive ion etching (RIE) for sharpening the tip, deep RIE for making structure thicknesses of the hundred-micrometer order, and wet undercut etching of the sacrificial layers for isolating the micromachine structures from the substrate. A general problem in the latter process is that the micromachine structures stick to the substrate due to the surface energy of the solvent. Novel technology has to be used to avoid sticking, as described in the next section.

7.4.2 Stick-Free Release of the Micromachined Structure from the Substrate

Thin film structures of the micromachine are liable to stick to the substrate during the drying process due to the surface energy of the solution. A photoresist-assisted release method to solve this problem has been successfully demonstrated [7]. In this method, the released structure is first fixed with a photoresist layer. After the release of the structures from the substrate, the fixing resist layers are removed by dry etching in an oxygen plasma ambient. The schematic cross-section and scanning electron micrograph (SEM

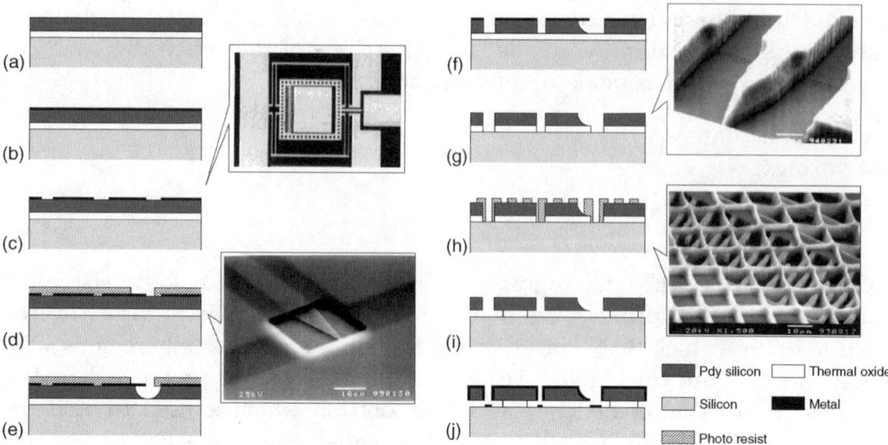

Fig. 7.7. Schematic cross-section and scanning electron micrograph of structures successfully released using the photoresist-assisted release method

micrograph) of the successfully released structures are shown in Fig. 7.7. This technology does not require any additional processing equipment and requires only a few extra processing steps to realize stick-free release. It thereby provides a practical solution to the sticking problem.

7.4.3 Dry Bulk Micromachined STM Chip Fabrication Process

Figure 7.8 shows a schematic representation of a bulk micromachined STM. The purpose of the device is to control the tunneling gap by a comb-drive actuator while the gap between two sharp probes is simultaneously observed by a transmission electron microscope. A through hole (Fig. 7.8b) is provided under the gap, allowing the electron beam to pass through the gap.

Fabrication of Facing Probes. The most critical part in fabricating the one-chip tunneling device is the preparation of sharp tunneling tips, which should be as sharp as 10 nm in diameter in order to confine the point of contact. We have tried two different approaches to tip-sharpening:

- etch-stop at the interface of boron-doped silicon,
- stress-induced oxidation of silicon.

In this section, we compare the results of these techniques.

It is known that wet etching of silicon is interrupted at the interface of heavily doped p-type silicon [13]. Boron (BF^{2+}) is ion-implanted at an acceleration voltage of 130 keV and a dose density of 3×10^{15} cm^{-2}, which corresponds to a boron concentration higher than 1×10^{20} cm^{-3} before annealing. As illustrated in Fig. 7.9a, a probe pattern of vertical side walls

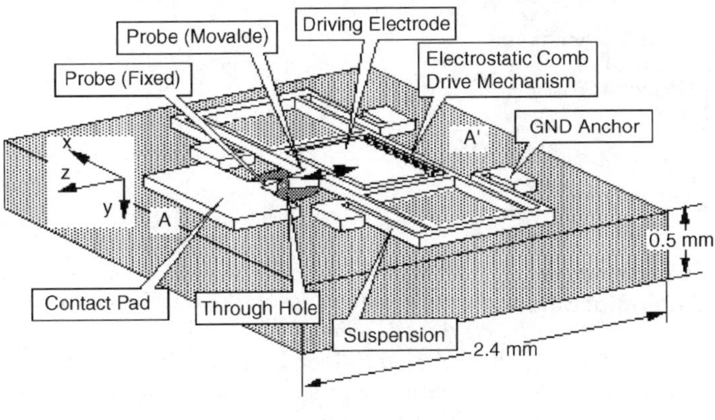

(a)

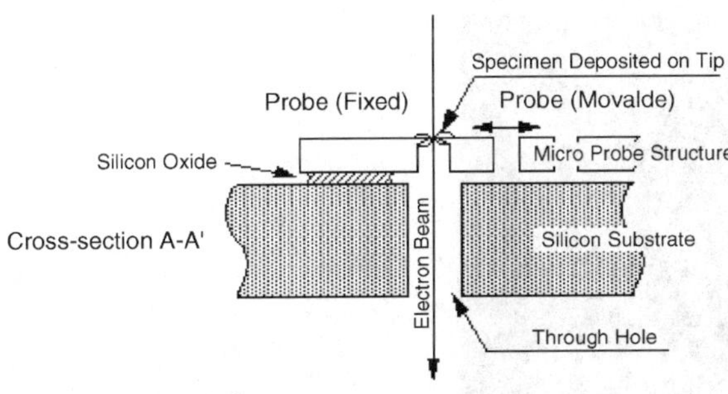

(b)

Fig. 7.8. Schematic diagram of one-chip tunneling probes and actuator. (**a**) An electrostatic comb-drive mechanism is used to control the tunneling gap. (**b**) Tunneling tips are suspended over a through hole for observation by the transmission electron microscope

is formed by inductively-coupled plasma reactive ion etching (ICP RIE) of 20 μm thick silicon. This is commonly known as deep RIE (DRIE). The structure is covered with a 2 μm thick silicon oxide layer formed by low-pressure chemical vapor deposition (LPCVD). After exposing the probe tip by stripping the silicon oxide, the silicon is etched in an aqueous solution of KOH (40 wt.%, 70°C) for 1 min. Since the top surface of silicon has been doped with boron, only the lower part of the probe is selectively etched, leaving thin eaves (see Fig. 7.9b).

Figure 7.10 shows a scanning electron microscope (SEM) view of the tunneling probe fabricated by the KOH tip-sharpening process. A pair of opposing tips is suspended over a through hole, with a separation of around 10 μm. In the close-up view (inset), we show a 100 nm thick tip of doped silicon extending out by 2 μm. Despite a sharp side-view profile, we found

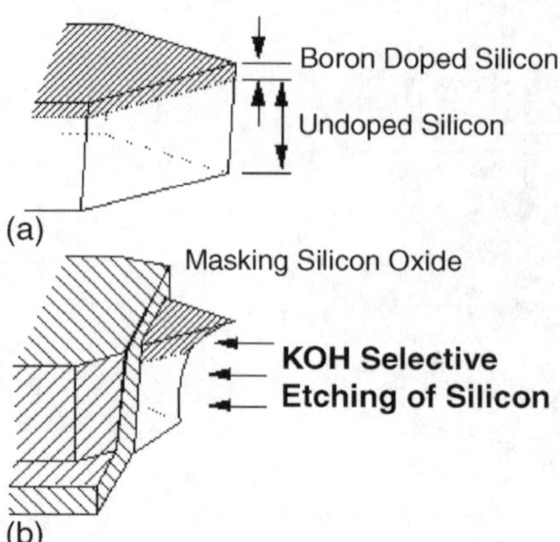

Fig. 7.9. Tip-sharpening process using etch-stop at boron-doped silicon. (**a**) Undoped silicon is covered with a layer of boron-doped silicon. (**b**) The boron-implanted region is not attacked by KOH aqueous solution

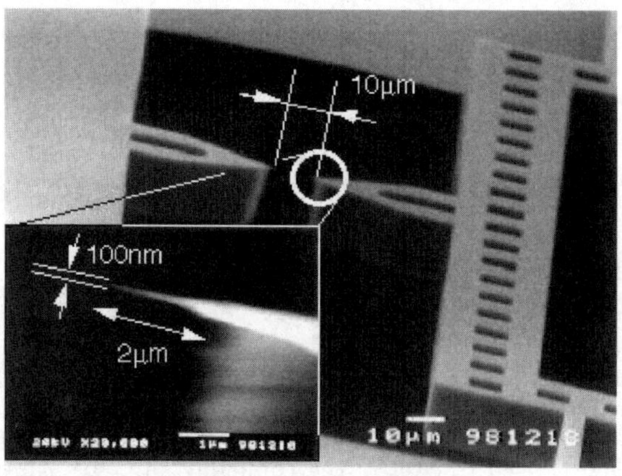

Fig. 7.10. Scanning electron micrograph of a tunneling tip made by the etch-stop technique with p^{++} silicon

some drawbacks to this technique. The sharpness of the top-view pattern was limited by the photolithographic resolution. For the same reason, the initial separation of the tips could not be made smaller than a micron. In addition to this, it was difficult to protect suspensions of high aspect ratio (5 μm wide and 20 μm high) with silicon oxide, and parts of them were attacked by KOH.

In contrast, tip sharpening by stress-induced thermal oxidation was satisfactory. This technique was developed by Itoh et al. [14,15]. They used electron beam (EB) lithography to produce a pair of tips and used them as a field emitter of electrons. We were able to obtain an equivalent result without EB lithography, using the fabrication process shown in Fig. 7.11. In Fig. 7.11a, a thick silicon layer is patterned as probes which are connected with a 2 μm waist on the photomask. The lower part of the waist is removed

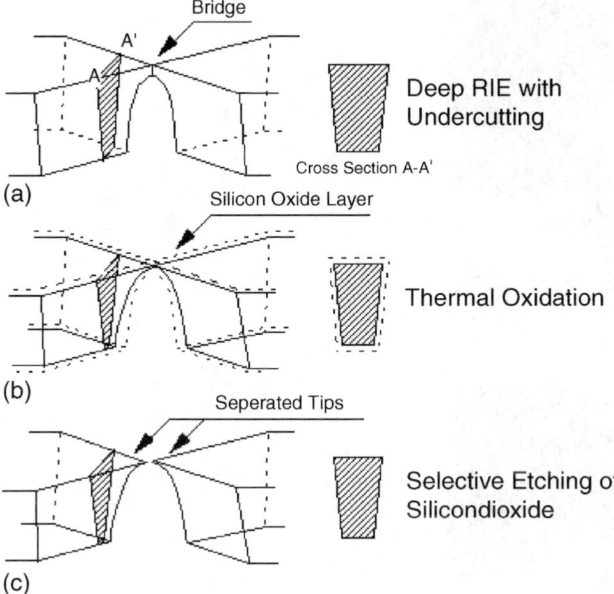

Fig. 7.11. Improved tip-sharpening process. (**a**) Two probes are produced, connected to each other by a small waist. The lower part is removed by the DRIE undercut effect. (**b**) The silicon inside the microbridge is separated into twin tips by stress-induced thermal oxidation. (**c**) Surface silicon oxide is removed by hydrofluoric acid to separate the tips

by the DRIE undercut effect. This produces a microbridge structure, typically 1 μm wide and 2 μm long. In Fig. 7.11b, the surface of the silicon structure is turned into silicon oxide by dry oxidation at the relatively low temperature of 950°C for 3 hr. Due to the stress induced by oxidation, the oxidation rate becomes slow at the bridge waist, resulting in a pair of sharp silicon cores remaining inside. Finally, in Fig. 7.11c, the surface silicon oxide is removed by hydrofluoric acid to separate the tips.

Figure 7.12a shows a silicon microbridge before thermal oxidation. The side-wall texture is due to the DRIE processing recipe on silicon. After oxidation and selective etching, the bridge is separated into a pair of tips with a gap of around 200 nm, as shown in Fig. 7.12b. The gap can be tailored by controlling the time of oxidation. By applying this technique, we think that it is possible to integrate more probes with apexes pointing in a sub-micron space. We have also demonstrated the formation of a silicon nanowire by the advanced anisotropic wet-etching process, which was developed by one of the authors [16,17]. The details of this process will be described later [18].

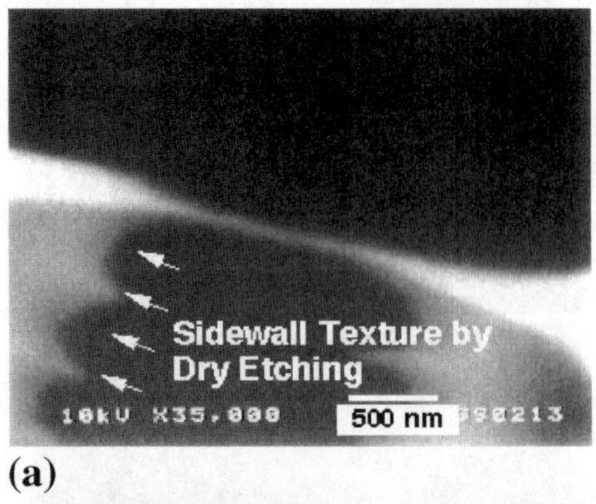

(a)

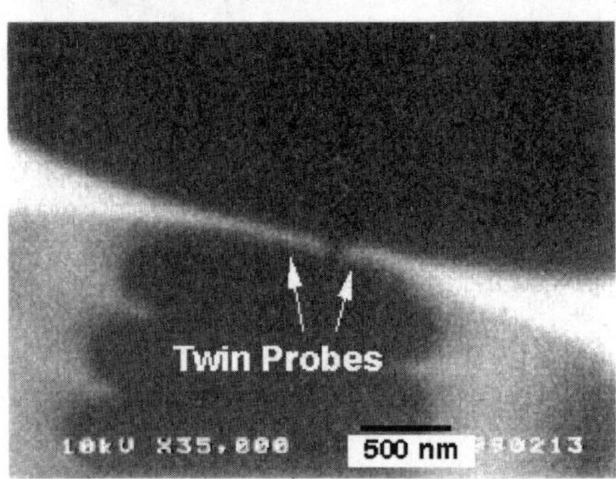

(b)

Fig. 7.12. Silicon microbridge (**a**) before and (**b**) after thermal oxidation. The two tips are separated by a 200 nm gap. The side-wall texture is due to the DRIE recipe

Total Process Sequence. Figure 7.13 shows the total fabrication process for the one-chip tunneling device based on tip sharpening by stress-induced thermal oxidation. There are five steps corresponding to Figs. 7.13a–e.

- We used a silicon-on-insulator (SOI) wafer which is laminated, from the bottom to the surface, in the following order: 525 μm silicon substrate, 1 μm silicon dioxide, and 20 μm thick single-crystalline silicon (see Fig. 7.13a).
- The rear side of the wafer is covered with a 1 μm thick silicon oxide layer formed by LPCVD. An etching mask, to be used in the next step, is formed by patterning the silicon oxide. The top silicon layer is etched into the microstructure by DRIE, using hard-baked positive-type photoresist of thickness 1 μm as etching mask. In this step, the microbridge structure is formed (see Fig. 7.13b).
- After protecting the front surface with a thick photoresist layer, a through hole (200 μm in diameter) is formed from the rear side of the substrate by DRIE. The etching reaction is interrupted at the interface of the buried silicon oxide layer. At the same time, 50 μm wide trenches are made around the device area (see Fig. 7.13c).

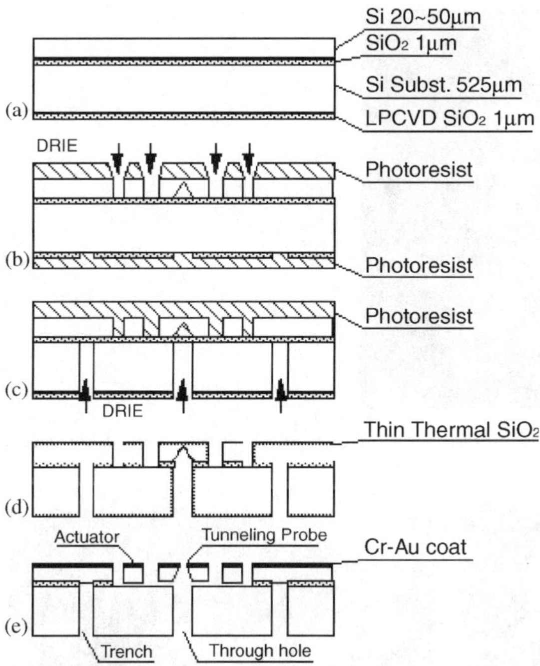

Fig. 7.13. Total fabrication sequence for the one-chip tunneling device based on stress-induced oxidation for tip sharpening. See text for details

- After removing the photoresist layers, a surface oxide of around 0.1 μm is grown by stress-induced oxidation at 950°C (see Fig. 7.13d).
- The microstructures are partially released from the base substrate by removing the buried silicon oxide with 47% hydrofluoric acid at room temperature. The etching time is controlled so that narrow parts, such as suspensions and frames, are released while large contact pads remain fixed to the substrate. Finally, the entire top surface is covered with 50 nm thick chromium and 200 nm thick gold layers (see Fig. 7.13e).

As shown in Fig. 7.14, the 2.4 mm × 2.4 mm chips are removed from the silicon substrate manually by breaking the supporting hinges indicated in the figure.

Figure 7.15a shows an SEM view of the wire-bonded chip, on which two identical tunneling control units are mounted. Contact pads are arranged on the chip for interconnection to the TEM sample holder. A close-up view of the actuator (before wire bonding) is shown in Fig. 7.15b. Two comb-drive mechanisms are connected in parallel to increase the total spring constant of the system. For the double-actuator structure in Fig. 7.15b, we substitute $w = 4$ μm, $h = 20$ μm, $L = 200$ μm, and $N = 2$, to obtain $k_{\mathrm{system}} = 89.6$ N/m, which is large enough to overcome the attractive atomic force.

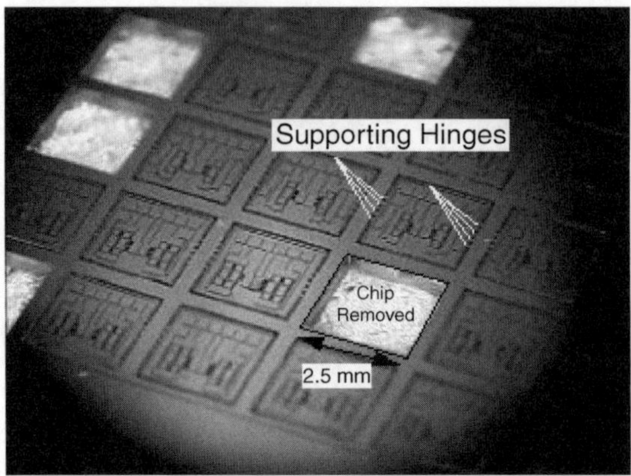

Fig. 7.14. Example of a processed silicon wafer from which some chips have been removed

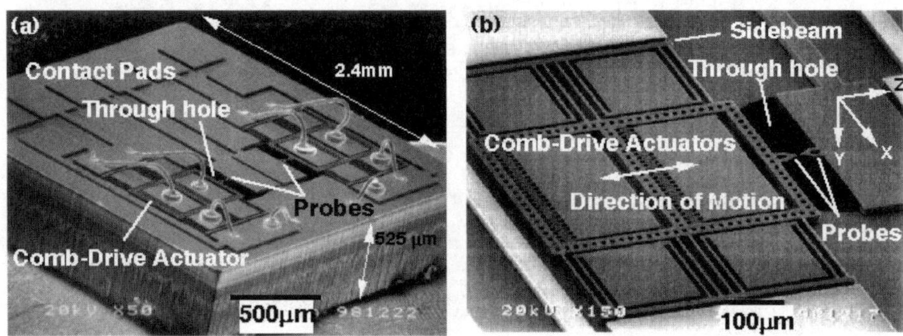

Fig. 7.15. (**a**) Scanning electron micrograph of a fabricated chip (dual package). The control voltage is applied through the gold bond wires. (**b**) Close-up view of the frame structure. Two comb-drive and suspension units are connected to increase the spring constant

7.4.4 Fabrication Process for Single-Crystal Silicon Nanowire and Nanoprobes

Nanoscopic probes can be realized by undercut etching as described in Sect. 7.4.1. However, the geometry of the probe is not well defined and is difficult to reproduce. Therefore, a more accurate way to fabricate nanoscopic probes has been developed. The probe is defined by the (111) crystal planes of single-crystal silicon.

Fabrication Process for Facing Nanowire Probes. Figure 7.16 shows the fabrication process for the nanowire. The silicon nanowire is fabricated by two-step anisotropic etching and local oxidation of silicon (LOCOS) on a SIMOX (separation by implanted oxygen) wafer [16]. Since the patterning is

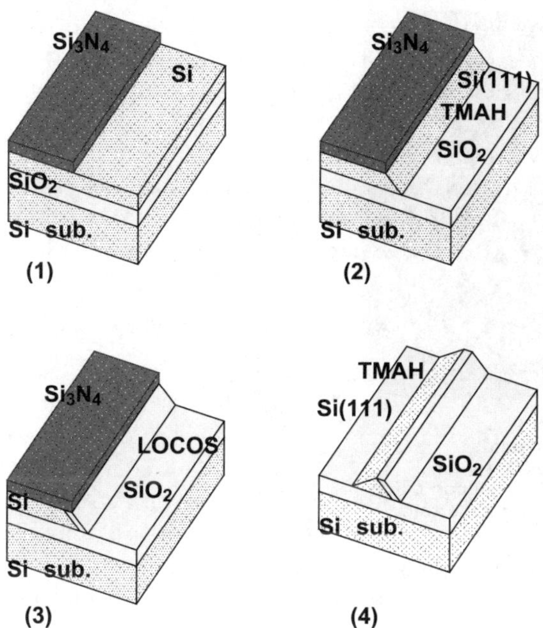

Fig. 7.16. Nanowire fabrication process

independent of the photomask resolution, the width of the wire can be made as narrow as the initial thickness of the silicon layer.

1. The starting material is a SIMOX–SOI (separation by implanted oxygen–silicon on insulator) wafer with a 60 nm layer of silicon and a 20 nm layer of silicon nitride deposited by LPCVD (low pressure chemical vapor deposition).
2. After patterning the silicon nitride mask into a stripe by RIE (reactive ion etching), the silicon layer is anisotropically etched in a TMAH (tetramethyl ammonium hydroxide) solution.
3. Local oxidation then covers the bevelled (111) side wall with a thin layer of silicon oxide (50 nm) and the silicon nitride mask is removed by RIE.
4. The exposed area of silicon is again etched in TMAH, and a silicon wire defined by two (111) surfaces is formed in the (110) direction. Note that the width of the wire is nearly equal to the initial thickness of the silicon SIMOX layer, being independent of the resolution on the photolithographic mask.

Once the freestanding nanowire has been made, it is cut into a pair of nanoprobes. The probe tips then face each other at a very short separation.

Results of Fabricating Facing Nanowire Probes. Figure 7.17 shows a SEM view of the fabricated nanowire probes, suspended from pads. Each

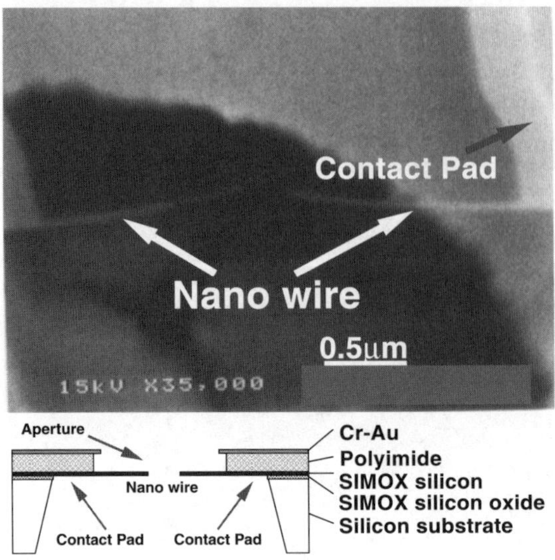

Fig. 7.17. SEM view of facing nanowire probes

probe is 55 nm wide, 39 nm high, and 1.5 μm long. This structure was de-

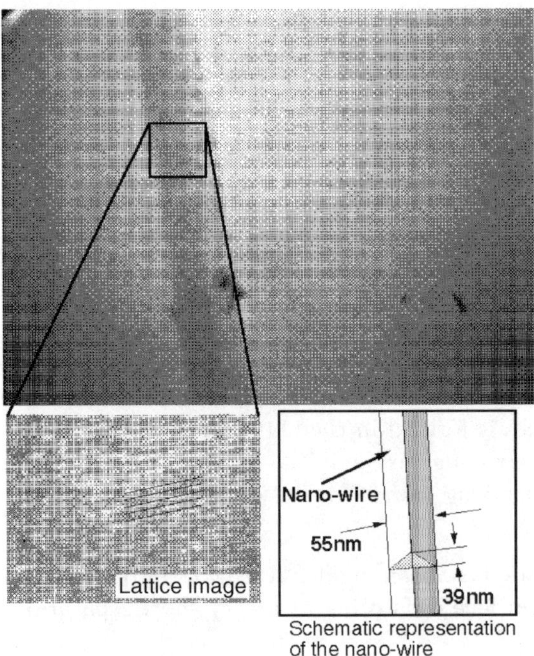

Fig. 7.18. TEM view of the nanowire. *Inset*: schematic representation

signed as a prototype to be loaded in the TEM chamber. To obtain high-resolution images, the entire surface has been covered with a thin polyimide layer (insulator) and chromium–gold (earthed shield). The TEM observation result is shown in Fig. 7.18. The lattice image of silicon was observed.

Fabrication Process for Twin Probes. The silicon nanowire process was also extended to realize nanoprobes that meet each other at an angle of 90°. The height and width of the probe are around 100 nm, while the separation between the tips is 100–500 nm, which can be further reduced by microactuators. The fabrication process is shown in Fig. 7.19. The major change with respect to the nanowire process is the shape of the mask pattern used to etch Si_3N_4 in the first process step (Fig. 7.19a). The second TMAH etching (Fig. 7.19d) is critical for precise adjustment of the tip spacing. The result is shown in Fig. 7.20.

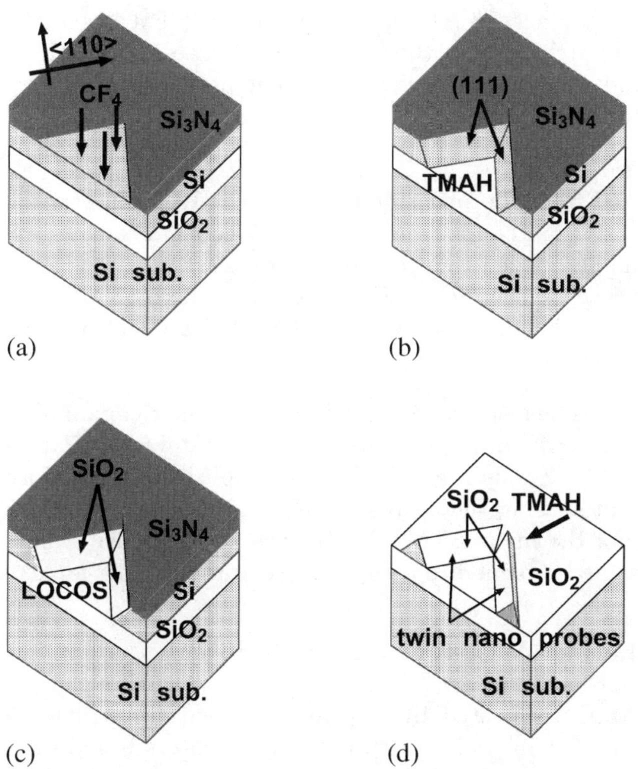

Fig. 7.19. Fabrication process for twin probes. (**a**) Patterning Si_3N_4 by RIE. (**b**) First anisotropic TMAH etching. (**c**) Local oxidation of silicon (LOCOS). (**d**) Second anisotropic TMAH etching

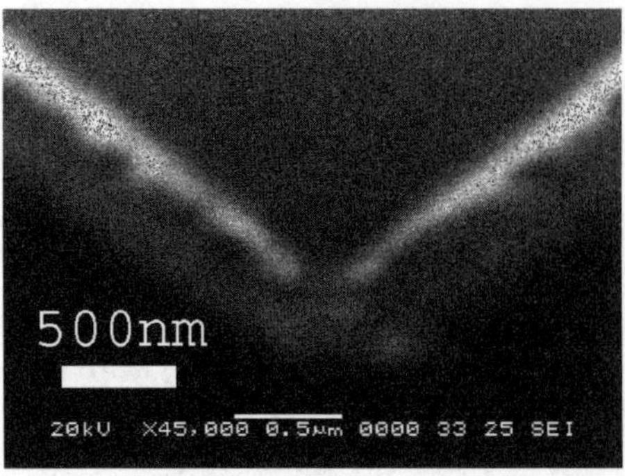

Fig. 7.20. SEM view of twin probes

7.4.5 Nanoprobes with Bulk Micromachined Actuators

Integrated Fabrication of High-Aspect-Ratio Comb-Drive Actuators with Nanowires. We combined the facing nanowire probes with a high-aspect-ratio comb-drive actuator to complete a micro-STM, as shown in Fig. 7.21. Anisotropic etching of a thin silicon layer, deep ICP RIE of the silicon substrate, and wafer bonding were used. The process is as follows (numbers correspond to Fig. 7.21):

1. nanowires are formed,
2. an aperture is formed from the rear side by deep ICP RIE (inductively coupled plasma reactive ion etching) for the TEM electron beam,
3. the silicon die is anodic-bonded to a recessed pyrex glass,
4. the top silicon is etched by deep ICP RIE to form the micro-STM structure. The die size is small enough (2.5 mm $\times$ 2.5 mm) to be loaded in the TEM chamber.

The actuator must have suspensions rigid enough to overcome the pulling-in force at the tunneling gap, which occasionally leads to unstable oscillation of the micro-STM. Therefore we used a 100 µm silicon substrate to produce a thick comb-drive actuator using the deep ICP RIE technique. Figure 7.22 shows a full SEM view of the micro-STM. Comb-drive actuators with more than 50 teeth were designed to generate a large electrostatic driving force.

Nanoprobes with Thermal Actuators. Thermally driven actuators were also used to drive nanoprobes. The actuator is driven by thermal expansion of a silicon structure which is heated by passing a current through it. A schematic representation of twin probes with thermal actuators is shown in Fig. 7.23. It should be noted that the device requires three silicon layers separated by two SiO_2 layers. A special substrate was fabricated by bonding two SOI substrates. The fabrication sequence for the device is shown in Fig. 7.24 and the final device in Fig. 7.25.

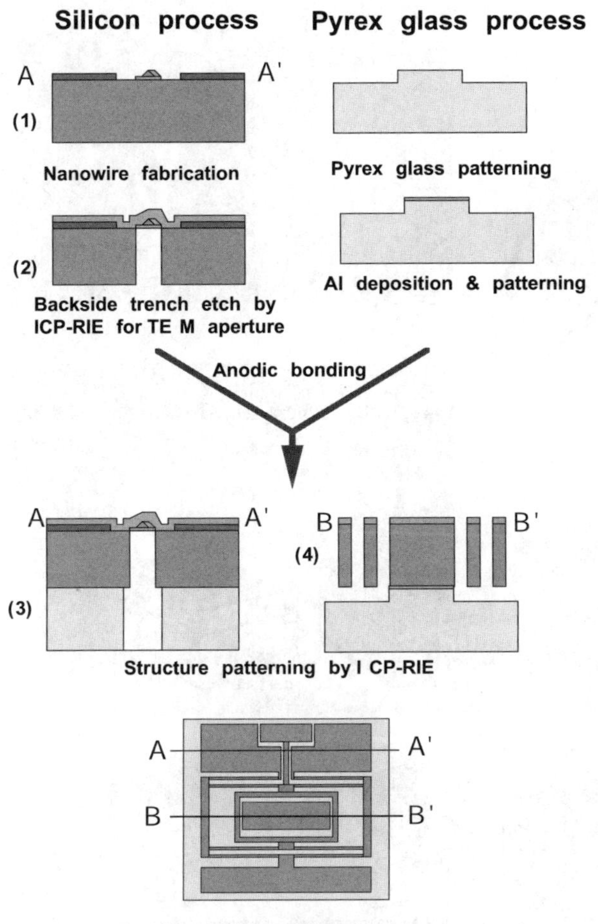

Fig. 7.21. Integrated fabrication of nanowires with a comb-drive actuator

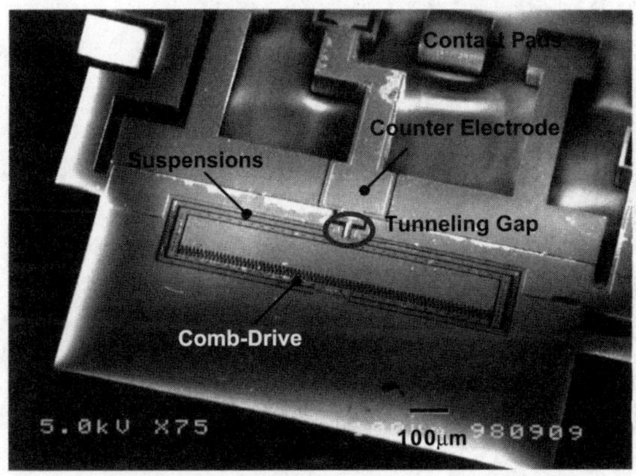

Fig. 7.22. SEM view of the micro-STM

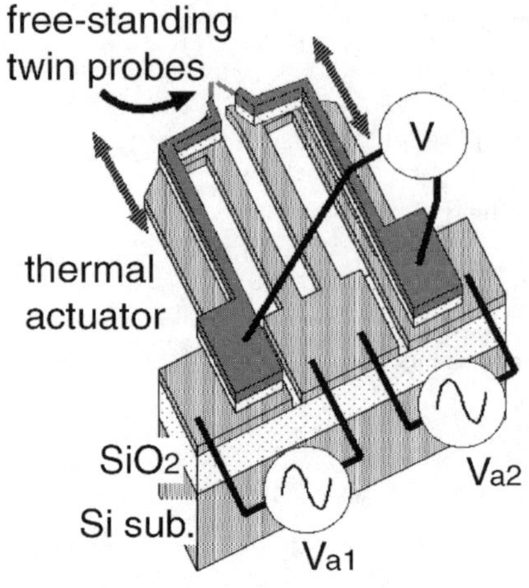

Fig. 7.23. Schematic drawing of the device

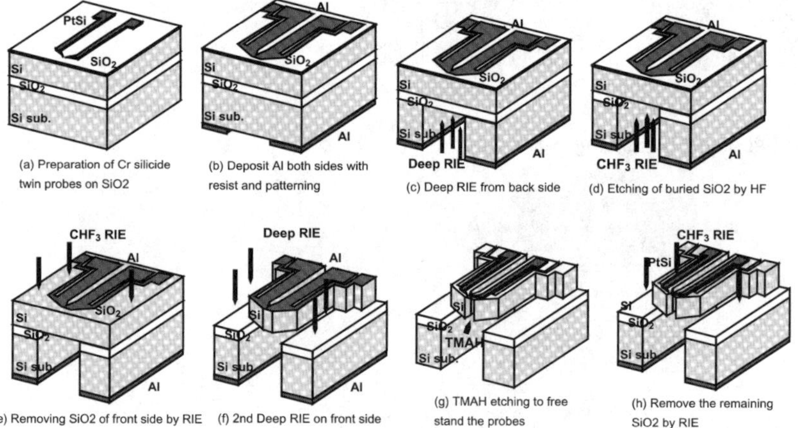

Fig. 7.24. Fabrication process for thermal actuator

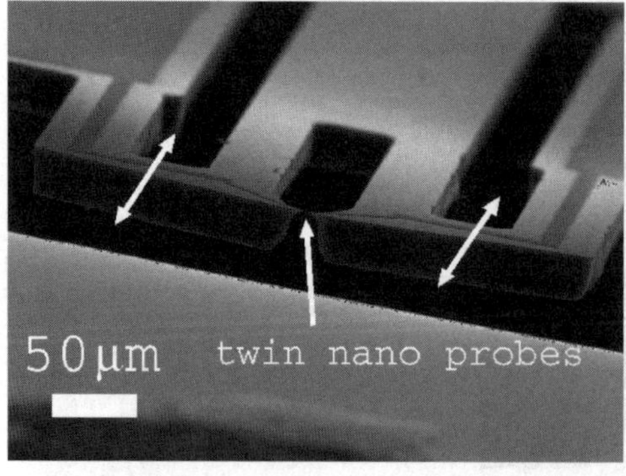

Fig. 7.25. SEM view of twin nanoprobes with thermal actuators

7.5 Characterization of the Fabricated Micro-STM

7.5.1 Operation in Air

The operating characteristics of the fabricated micro-STM were examined [6]. The micro-STM setup is shown schematically in Fig. 7.26a, where the micro-STM is placed against an opposing electrode connected to a piezo actuator. The motion of the electrode driven by the piezo actuator is compensated by the electrostatic actuator through the tunnel current feedback. Figure 7.26b is a SEM micrograph of the micro-STM used in this measurement. The tunnel current was measured against time, while the piezo actuator bias was changed in a triangular wave manner, and the movement of the opposing electrode was compensated by the electrostatic actuator, as shown in Fig. 7.27. The figure clearly indicates that the tunneling current was kept almost constant, while the electrostatic actuator compensated the piezo actuator motion by moving the tip by about 80 nm. These results imply that the electrostatic actuator perfectly controls the tip motion by using the feedback control of the micro-STM system. The characteristics of the electrostatic actuator were thereby ascertained experimentally.

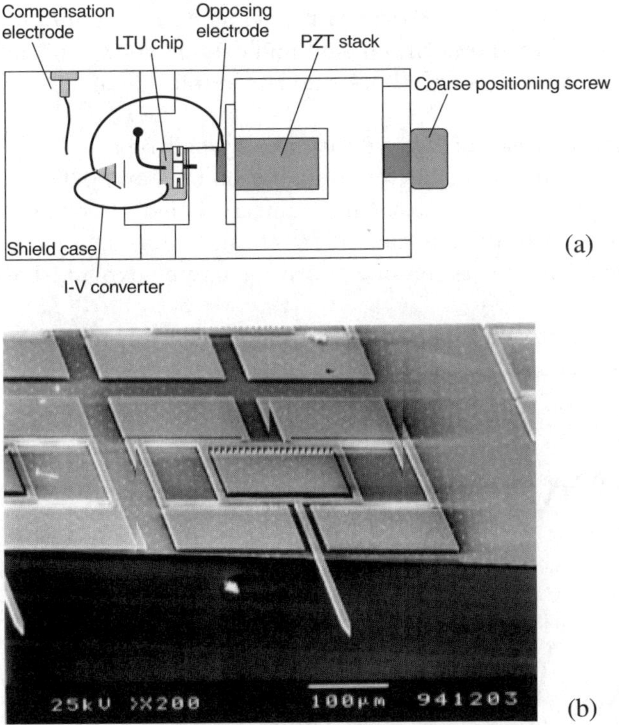

Fig. 7.26. (**a**) Schematic figure and (**b**) scanning electron micrograph of the setup for measuring the operating characteristics of the fabricated micro-STM

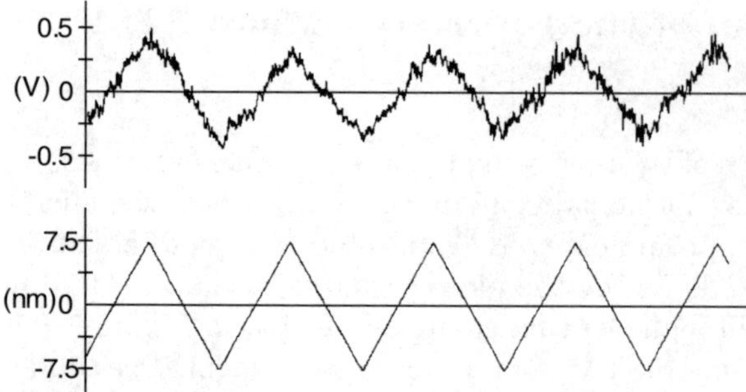

Fig. 7.27. *Upper*: microactuator control voltage. *Lower*: displacement generated by PZT. Tunneling current 1 nA, tunnel bias 110 mV, frequency 10 Hz, actuator bias 64 V

7.5.2 Operation in Vacuum

Operational Instability in Ultrahigh Vacuum. A micro-STM was operated in ultrahigh vacuum (UHV) in order to understand the mechanism in the absence of contaminants. A micro-STM chip was placed in a vacuum chamber and heated to about 200°C with the chamber. The chamber was pumped down to the order of 10^{-11} torr. The micro-STM tunneling current exhibited a complex vibration. Figure 7.28 shows the current and driving voltage of the micro-STM. The reference value was DC 10 nA and the amplitude of current fluctuation was 1 decade.

The current waveform consists of two vibrations: a small one with frequency about 15 kHz and a large one with period between 0.25 and 0.32 ms. The former vibration frequency is close to the mechanical resonant frequency. By assuming that the large vibration is caused by atomic forces in the attractive region, the mechanism of this complex vibration can be explained as follows.

Suppose that the tip approaches the opposing electrode. When the tip comes to the point A in Fig. 7.29, the slope of the attractive atomic force

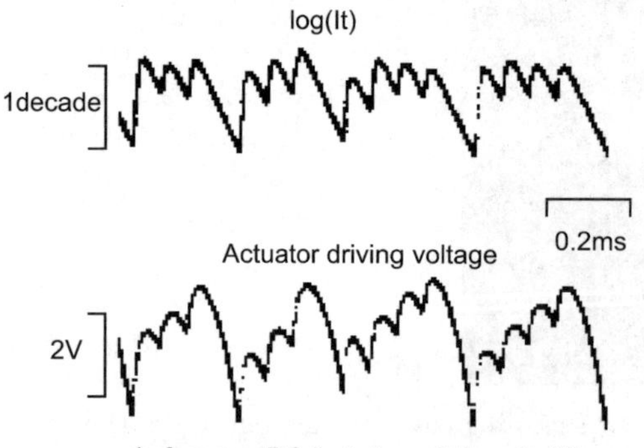

Fig. 7.28. Vibrations in the current and driving voltage of the micro-STM in UHV

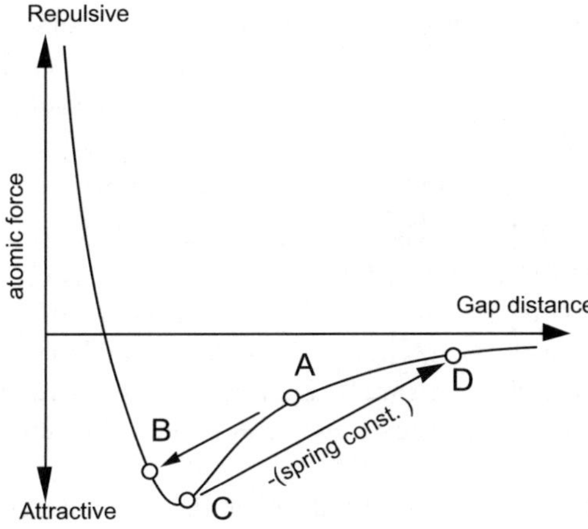

Fig. 7.29. Atomic force and tip jumping

exceeds the spring constant of the actuator. The result is that the tip is pulled to the point B. The tunneling gap reduces further, although the tunneling current is larger than the reference, and the tip starts vibrating because of the inertial force gained in the jump from A to B. The driving voltage slowly reduces to retract the tip. When the tip comes to point C, the spring constant exceeds the slope of the attractive atomic force and the tip jumps back to point D. The driving voltage increases and the tip jumps backs again from A to B. The small vibration helps the jumping to take place. The waveform resembles a chaotic vibration.

Let us estimate the amplitude of the atomic force variation across the tip jump. The depth of the dip in the driving voltage is about 2 V. The DC component of the driving voltage is about 100 V and the initial gap about 2 μm. The displacement sensitivity is calculated to be 4×10^{-8} m/V from (7.2). Multiplying the spring constant 2.2 N/m by this value, the force per unit driving voltage generated by the actuator is obtained as 4.4×10^{-7} N. Consequently, the force amplitude is found to be 8.8×10^{-7} N. Considering that the generating force of the actuator is of the order of 10^{-6} N, this force amplitude is too large a disturbance for the surface micromachined actuator. In any case, we cannot determine the reason for the difference between the experiments in the air and in UHV without stabilizing the control system.

Increased Spring Constant and Force. We designed and fabricated a micro-STM with a spring constant of about 100 N/m and a minimum increase in dimensions, especially in the width of the unit. The high-rigidity micro-STM has a train of 10 identical actuators to increase the spring constant and the generating force by a factor of ten. This modification does not reduce side-by-side integration density. The rest of the increase in the spring constant was achieved by increasing the width of the suspension beams. In order to increase the generating force per pair of comb teeth, the tooth shape was changed from straight to step-shaped [19]. This increases the parts that generate electrostatic force and reduces the gap between the teeth. The total spring constant then increased to 126 N/m.

High-Rigidity Micro-STM in UHV. The high-rigidity micro-STM was put in UHV. Controller gains were adapted to the new mechanical parameters. We found the range of the tunneling bias and tunneling current in which the high-rigidity micro-STM can be stabilized. Figure 7.30 shows the actuator driving voltage and the logarithm of the tunneling current. The current swung from 1.3 to 5.6 nA. The figure was averaged over 256 results, because the waveform was noisy.

From the initial gap 1.5 μm and the DC component of the driving voltage 100 V, the actuator displacement sensitivity is calculated to be 30 nm/V. Since the amplitude of the driving voltage is 120 mV, the displacement is estimated to be 3.6 nm. On the other hand, the tunneling current variation indicates a displacement less than 0.1 nm. This is the same disagreement as we found in the air, although the environment was UHV and no elastic substance can exist between the tip and the opposing electrode. Therefore, it is natural to think that the source of the gradient force is atomic. Multiplying the spring constant 126 N/m by the displacement sensitivity 30 nm/V, the force sensitivity is calculated to be 3.8×10^{-6} N/V. Consequently, the force amplitude which corresponds to the variation of the current (1.3–5.6 nA) is obtained as

$$(3.8 \times 10^{-6}\ \mathrm{N/V}) \times 120\ \mathrm{mV} = 4.5 \times 10^{-7}\ \mathrm{N}\,.$$

In reference [20], the tunneling current and force were measured simultaneously by AFM. The force amplitude calculated above corresponds well to a plot of the current and the force in the reference.

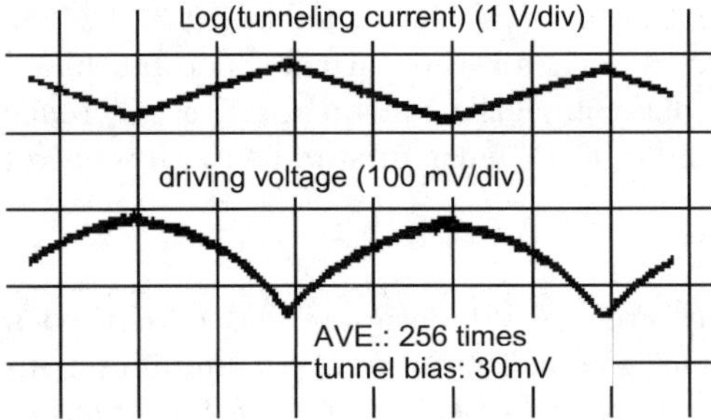

Fig. 7.30. Waveforms from a high-rigidity micro-STM in UHV

7.6 Possible Applications of Micromachine STM Technology

7.6.1 Micromachine STM for Sub-100 nm Lithography System

Potential and Problems of SPM Lithography. Scanning tunneling microscope (STM) was invented in 1982 [1] and has been used not only to observe surfaces on the atomic scale, but also to manipulate atoms on the surfaces [21,22]. It was also demonstrated that a scanning probe microscope (SPM) can delineate nanometer scale patterns on surfaces by oxidizing the metal and silicon layers [23,24], by mechanically scratching the thin films [25], and by exposing self-assembled monolayers [26], but very few attempts were made to form patterns on resist layers [27]. The principle of exposure by STM is that electrons emitted from the SPM tip release energy from the exposed top layer and transform it by mechanisms such as oxidation, dissociation and cross-linking, except in the mechanical method, in which mechanical deformation of the patterned layer takes place. Therefore, the width of the fabricated pattern depends on such parameters as the exposed top layer material, the thickness of the top layer, the electron exposure dose, the exposing acceleration voltage and the tip structure.

The most conventional method for fabricating sub-100 nm patterns using SPM is the oxidation of metal/semiconductor surface layers [23,24]. The oxidized layers function as etching masks against the underlying layers. A

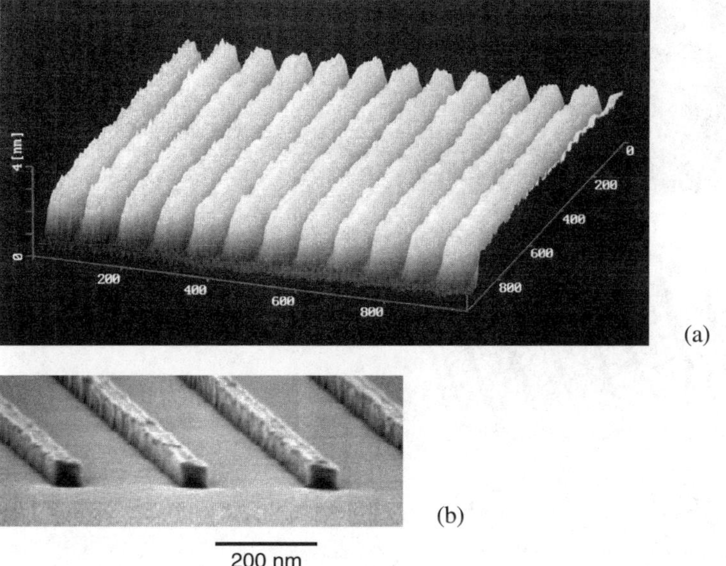

Fig. 7.31. 50 nm line and space patterns delineated by (**a**) oxidation of silicon and (**b**) direct exposure of thin resist layers

typical example is shown in Fig. 7.31a, where 50 nm line and space patterns were delineated by oxidation of silicon [28]. Patterns can also be transferred to organic materials, and 50 nm line and space patterns formed by direct exposure of thin resist layers are also shown in Fig. 7.31b [27]. In the resist exposure method, the pattern size depends on such parameters as resist thickness, exposure dose and acceleration voltage, and the minimum pattern size formed using this method was less than 30 nm. These results clearly indicate the potential for SPM lithography in the sub-100 nm lithography regime.

The ultimate lithography technology using SPM would be atomic-scale delineation of patterns. The first example of an atomic line fabricated on the hydrogen passivated silicon (100) surface was reported in [29]. Line patterns between the atomic scale and 10 nm wide were formed on a silicon substrate by removing hydrogen atoms. This was done by increasing the tunneling current and/or the electric field applied from the STM tip. The technology was

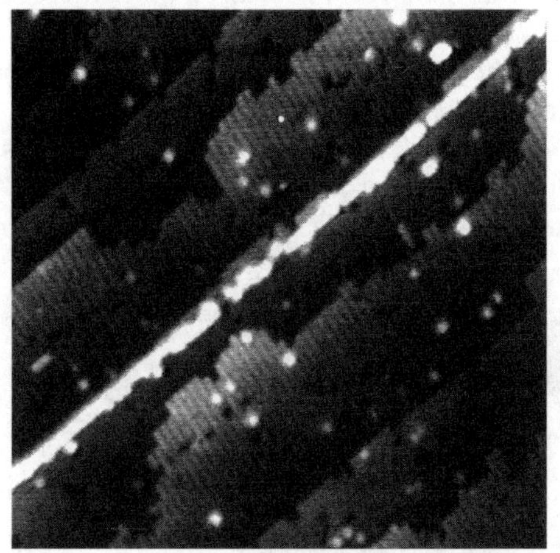

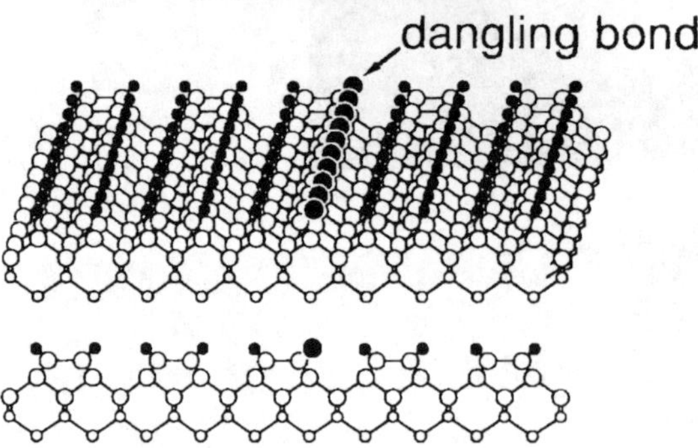

Fig. 7.32. A single dangling-bond wire successfully fabricated by STM on a hydrogen passivated silicon (100) substrate

successfully replicated, and even a single dangling-bond wire was successfully fabricated on the hydrogen passivated silicon (100) substrate, as shown in Fig. 7.32 [30]. The figure shows a schematic bird's eye view and cross-section of the hydrogen passivated silicon layer as well as STM images of the corresponding dangling bond wire structures. These results clearly indicate that an atomic scale line less than 1 nm wide can be fabricated on silicon using SPM, and that 'atom electronics' [31,32] could become a reality by extending this technology.

However, SPM lithography is quite limited with regard to writing speed. It is reasonable to assume that the writing speed is determined by the minimum pattern area (A_{min}) multiplied by the maximum repetition rate, which is equal to the intrinsic vibration frequency of the SPM unit (f_{IV}). Therefore, the maximum area of pattern transfer (A_{P}) per unit time is given by

$$A_{\mathrm{P}} = A_{\mathrm{min}} \times f_{\mathrm{IV}} = l_{\mathrm{min}}^2 \times f_{\mathrm{IV}} \;, \qquad (7.4)$$

where l_{min} denotes the minimum line width of the delineating patterns. If l_{min} and f_{IV} are 100 nm and 1 MHz, respectively, A_{P} is equal to 10^{-4} cm^2/s. Figure 7.33 compares the minimum pattern size and writing speed of various lithography systems, including optical, X-ray, electron beam and SPM [33]. As indicated in Fig. 7.33, a practical system has to have a writing speed greater than 1 cm^2/s. However, equation (7.4) suggests that the fastest writing speed of a single SPM unit is about four orders of magnitude lower. The only solution would be a parallel lithography system [34] using micromachine

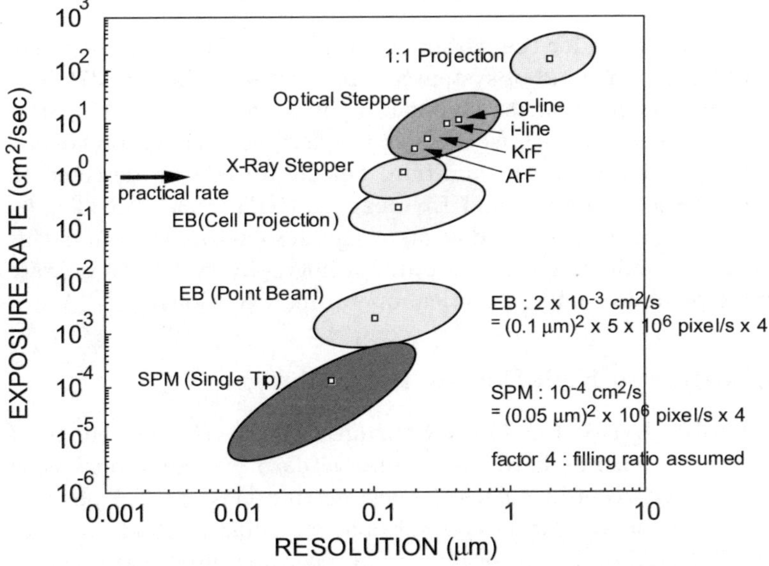

Fig. 7.33. Comparison between the minimum pattern size and writing speed of various lithogaphy tools, including optical, X-ray, electron beam and SPM

technology [35,36]. The following sections describe the possible composition of a practical lithography system.

Advantages of the Micro-STM Unit and STM Parallel Lithography System. The micro-STM has the following advantages for advanced lithography, since it is very small, can be operated in parallel by the thousands and is fabricated using advanced ULSI technologies [12]:

- a very fast writing speed as high as 1 MHz,
- a very high packing density of more than 10^2–10^4 devices/mm^2,
- more than 10^4 devices can operate in parallel by integrating controller circuits on the same chip,
- low operating voltages of around 30 V, reduced risk of cross-talk, and no need for special high-voltage circuit/device structures.

Consequently, micro-STM-based parallel lithography systems are a quite promising candidate for sub-100 nm applications.

The full wafer parallel lithography system using micro-SPM units is shown schematically in Fig. 7.34. The full wafer complex system is made up of plural writing units which consist of many SPM probes. The number of SPM probes to be integrated in the full wafer complex would lie somewhere between 10^3 and 10^5, depending on the minimum resolution and the required writing speed, as shown in Fig. 7.33 [33,34]. About 104 SPM probes would be required to fabricate a 50 nm device at the mass production writing speed of 1 cm^2/s, while about 10^5 probes would be needed if the device size was shrunk down to 10 nm.

The reliability problem for the parallel micromachine lithography system would be solved using the kind of system architecture described in [37]. The detection system embedded on the chip can monitor the failure of the SPM unit in real time, observing such features as tip sharpness, signal processing error and even position error. In addition, tip failure can be detected in situ using the needle formation and tip imaging (NFTI) method [38], by which the tip apex can be delineated at an atomic resolution. Therefore, this monitoring system would enable the parallel lithography system to operate with reasonable reliability under practical operating conditions.

7.6.2 Application to High-Density Data Storage

Tunneling Probe Array for Data Storage. The recording density of data storage devices such as hard disk magnetic data storage is increasing very rapidly. Current recording densities are getting close to the physical limits for the achievable density of conventional recording methods, such as magnetic recording and optical recording. SPM recording represents one possible breakthrough to ultrahigh-density data storage [39–41]. The drawback

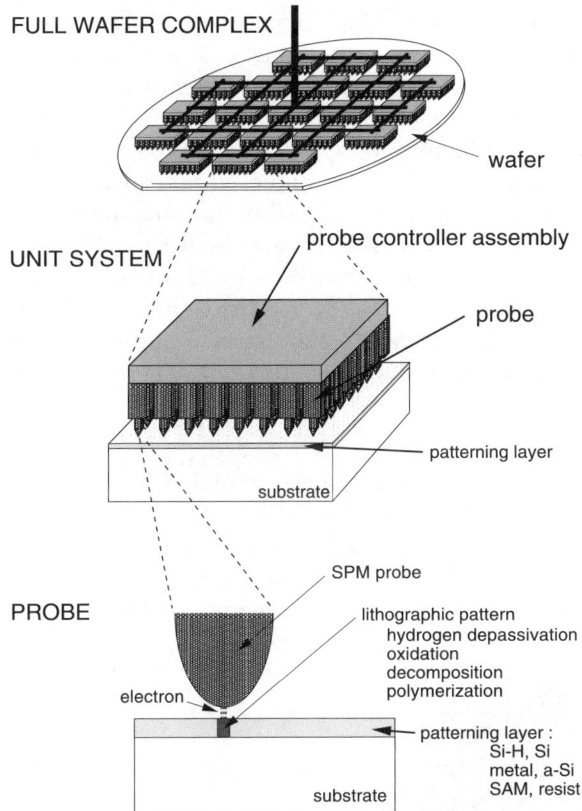

Fig. 7.34. Schematic illustration of the full wafer writing system using μ-SPM

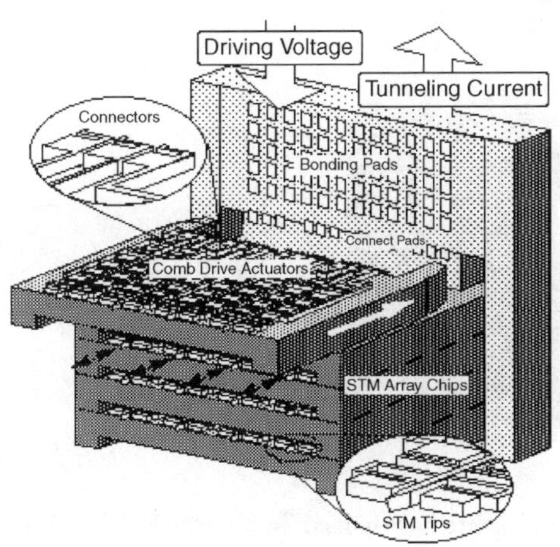

Fig. 7.35. Stacked array of tunneling probes for data storage applications

with SPM data storage is the slow writing/reading speed. In order to overcome this difficulty, we are developing a read/write head for atomic storage with a matrix of tunneling probe tips which extend out of the chip as shown in Fig. 7.35. Each chip has an array of probes whose z-positioning is controlled by the electrostatic comb-drive mechanism. The probe matrix is positioned on a surface in the vicinity of the tunneling current and used to read and write atomic-scale structures. Coarse positioning is achieved by the tube piezoelectric actuator on which the probe matrix is loaded, and each tunneling gap is controlled by the integrated microactuator over a range of a few microns. Driving voltages are supplied from the external source through contact pads and pins. Tunneling currents are transferred to the external amplifiers. We plan to improve the data transfer speed of atomic storage by increasing the number of channels.

Fabrication of Probe Array. Figure 7.36a shows a full SEM view of a tunneling probe array of 20 µm thick silicon-on-insulator (SOI). The silicon has been patterned by dry etching using ICP RIE with a photoresist mask. The probe is suspended with a set of folded suspensions and free to move along its axis. Three chains of comb-drive actuators are arranged in-between

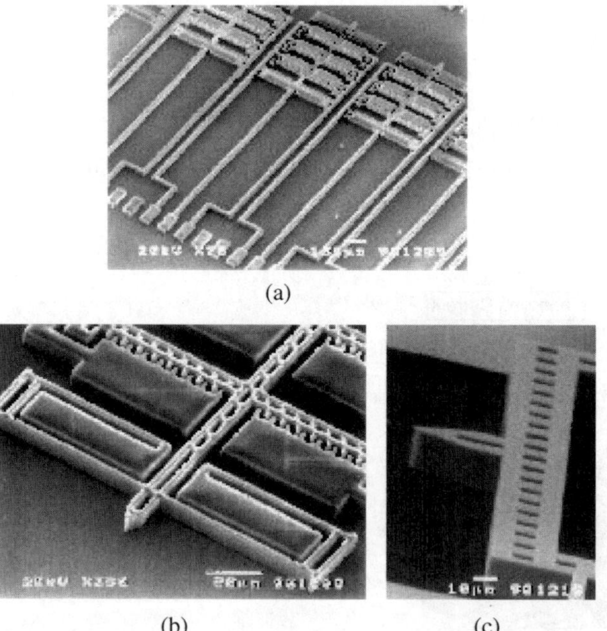

(a)

(b) (c)

Fig. 7.36. SEM views of tunneling probe units fabricated by ICP RIE. (**a**) Full view with contact pins. (**b**) Close-up view of the tunneling probe. (**c**) Sharp tip formed by KOH selective etching of heavily boron-doped silicon

the suspensions. The electrode pattern on the actuators can be traced down to the contact pin at the bottom of the photograph. Each contact pin is composed of a cantilever with a tiny contact nail, which will be placed in the contact pad on the base substrate.

The probe tip was sharpened by selective etching of doped silicon. The SOI wafer used was first doped with implanted boron ions to over 10^{20} cm^{-3} and patterned by ICP RIE. The whole surface was then covered with a 3 μm thick silicon oxide layer by LPCVD (low pressure chemical vapor deposition) and a window was opened just at the probe tip to expose the probe to a KOH solution at 80°C. In this wet etching, the undoped silicon body is etched and the doped silicon layer remains. The tip is thus sharpened in thickness as shown in Fig. 7.36c. The structure was oxidized in 1 000°C O_2 ambient gas to form silicon oxide on the surface. The oxidation process is stress-sensitive, and the oxidation rate is slower at the sharp corner. Therefore, one can obtain a sharper tip after stripping the oxide layer. The subsequent process and assembly procedure is under investigation.

7.6.3 Experimental Tool for Understanding Basic Physics

Observation of the Tunneling Gap. The micro-STM was used as a sample for the transmission electron microscope (TEM) and the vacuum tunneling gap between the tip and sample during the tunneling regime was directly observed for the first time [9] since the invention of the STM in 1982 [1]. This achievement also confirms the possibility of controlling the tip apex shape for reliable lithography and atom manipulation. In addition, detailed analysis of the interaction between the tip apex and the sample should lead to a deeper understanding of the atom/molecule manipulation mechanism, which should also help to control patterning of the layers.

The micro-STM chip was inserted into the TEM using a modified HF-2000 (Hitachi) TEM side-entry sample holder. Figure 7.37 shows a TEM micrograph of the vacuum tunneling gap between the micro-STM tip and sample, where the voltage and current across the two electrodes are 1.4 V and 4 nA, respectively. The Au(200) lattice images were observed. This sets the length scale standard for the TEM image, and the width of the vacuum tunnel gap between the tip and the sample measures about 1 nm. An accurate measurement of the voltage–current/gap distance relationship should make experimental evaluation of the vacuum tunneling theory possible.

The results shown in Fig. 7.37 clearly indicate that further improvements in TEM resolution and micro-STM performance should make it possible to investigate single atoms and molecules around the STM tip and sample region. Furthermore, they suggest that material transport between the two electrodes will also be observable. For example, the transport of a gold atom or cluster between the tip and sample, or the dependence of the material transport direction on the bias polarity can be observed as functions of tunneling bias and current. Hence, a detailed analysis of the mass transport phenomenon

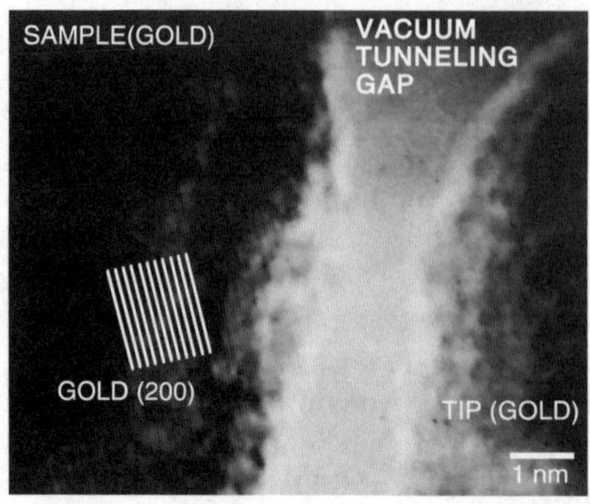

Fig. 7.37. Transmission electron micrograph of the tunneling gap between the micro-STM tip and sample

should lead to a better understanding of the mechanisms of atom manipulation by STM. This in turn will lead to the establishment of a reliable micro-STM parallel lithography technology. The program is now in progress with the support of the Japan Science and Technology Corporation (JST) CREST program to improve the resolution of the TEM and the performance of the micro-STM.

Direct Visualization and Control of Nanoscopic Physical Phenomena. In the vacuum tunneling gap, the electric field strength reaches 10 MV/cm and the current density exceeds 100 MA/cm^2. These are localized in a region of atomic size near the tip and their magnitudes change significantly within a fraction of a nanometer. Various phenomena such as atomic/electronic transport, localized excitation and relaxation, and chemical reactions between single molecules occur in such an extreme environment. The project aims to visualize the behavior of an individual atom or molecule in the tunneling gap and to analyze their physical properties. The three major research activities in the project are:

- development of a micromachined STM chip for controlling the electric field in the gap with atomic level precision,
- development of a phase-detection electron microscope [42] and micro CT imaging for visualizing individual atoms, a single molecule or the electric field distribution in the gap,
- first principles theoretical calculation of physical phenomena in the tunneling gap.

The micromachined STM will be operated in the ultra-high resolution microscope. The physical phenomena in the tunneling gap will be observed visually. Our approach is to set the facing nanowire probes in one of the split paths of a TEM electron beam to obtain images by electron interference, as shown in Fig. 7.38. The electron beam goes through the tunneling gap of the micro-STM and projects a transmission image by electron interference.

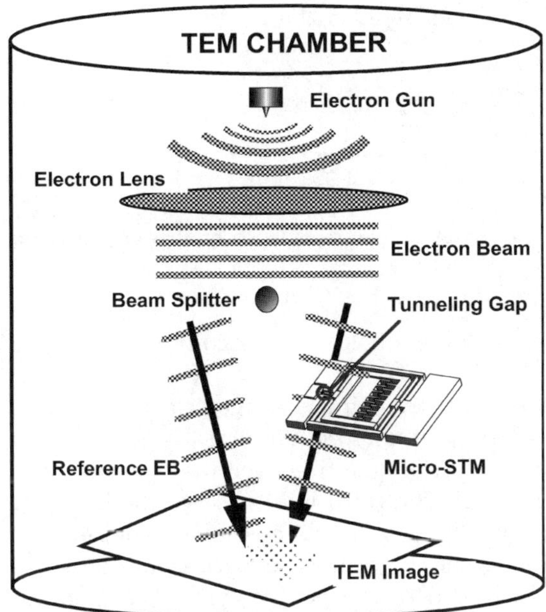

Fig. 7.38. Micromachined STM in an electron interferometric TEM

The micro-STM is composed of two elements: an electrostatic microactuator with a high aspect ratio (the ratio of height to width) and a nanowire tunneling probe (Fig. 7.39). A comb-drive actuator with high aspect ratio is used to generate a large driving force and to overcome the pull-in attraction force at the tunneling interface. A nanowire defined by (111) crystal planes is

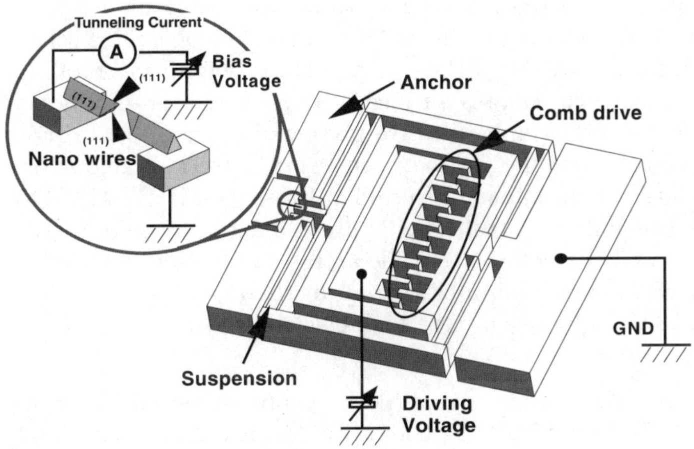

Fig. 7.39. Schematic representation of the micro-STM

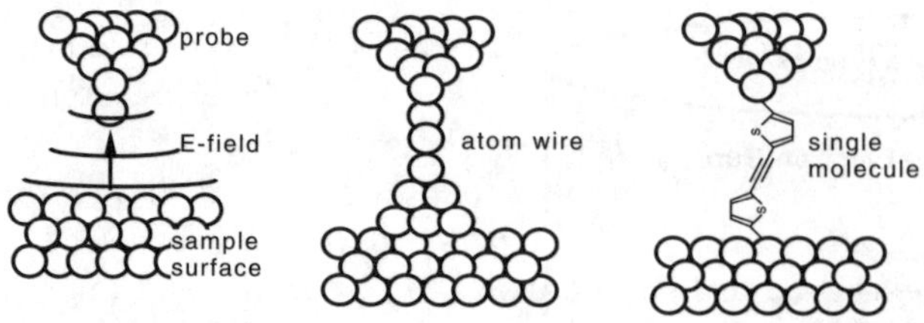

Fig. 7.40. Direct visualization of the tunneling gap reveals many physical phenomena

formed at the micro-STM probe. A through hole is prepared in the substrate as an EB aperture.

As shown in Fig. 7.40, there are three primary targets in the project. The first is simultaneous obervation of atomic motion and 3D electric field distribution. The second is measurement of quantized electron transport in a nanoscopic wire by direct observation. The third is to observe the behavior of a single molecule in the gap and to measure its mechanical and electrical characteristics. In the final stage of the project, experimental data will be analysed theoretically. This will lead to the precise control of quantum phenomena in the extreme environment of the tunneling gap.

7.7 Conclusion

This chapter was concerned with the current status of Japanese research and development into micromachine scanning tunneling microscopes (micro-STMs). The design principles, fabrication technologies and operating characteristics of representative research results were summarized. We also described the possible application of micro-STM technology to future nanolithography tools in the sub-100 nm regime, to arrayed SPM data storage devices, and to scientific research in nanoscience and technology. SPM lithography systems, which use 10^3–10^5 SPMs as a pattern delineation tool, are one of the most promising candidates for massively parallel mass-production tools in the sub-100 nm lithography era. We presented observational results concerning the vacuum tunneling gap between two micro-STM conductors, using a high resolution transmission electron microscope (HRTEM). This demonstrated the possibility of realizing highly reliable atomic resolution lithography systems and characterizing single atoms and molecules for scientific research. The STM tunnel gap observation method can be extended to characterize atoms and molecules in very high electric fields.

Acknowledgments. The authors would like to express their sincere gratitude to Prof. Hiroshi Toshiyoshi, Mr. Makoto Mita, Mr. Kunihoro Kakushima, Dr. Akira Endo, Dr. Shinji Okazaki, Dr. Tsuneo Terasawa, Dr. Toshiyuki Yoshimura, Dr. Tokuo Kure, Dr. Mark I. Lutwyche, Mr. Masayoshi Ishibashi and Dr. Tomihiro Hashizume for fruitful discussions and suggestions. One type of micro-STM chip was fabricated by the ULSI Research Department and Advanced Device Department, both at the Hitachi Central Research Laboratory.

References

1. G. Binnig, H. Rohrer, Ch. Gerber, E. Weibel: Phys. Rev. Lett. **49**, 57–60 (1982)
2. T.W. Kenny et al.: *Wide Bandwidth Electromechanical Actuators for Tunneling Displacement Transducers*, Technical digest of ASME/IEEE J. of Microelectromechanical Systems **3**, 97–104 (Napa Valley, CA, U.S.A. 1994) pp. 192–196
3. S. Akamine et al.: *A Planar Process for Microfabrication of a Scanning Tunneling Microscope*, Sensors & Actuators A: Physical **23**(1)–(3), 964–970 (1990)
4. Y. Xu, S.A. Miller, N. MacDonald: *Microelectromechanical Scanning Tunneling Microscope*, Tech. Digest 8th International Conference on Solid-State Sensors and Actuators (Transducers '95), Stockholm, Sweden, June 25–29 (1995) pp. 640–643
5. D. Kobayashi, H. Fujita: Trans. IEE Japan **116** E, 297–302 (1996)
6. D. Kobayashi, H. Fujita: Trans. IEE Japan **116** E, 339–344 (1996)
7. D. Kobayashi, C.J. Kim, H. Fujita: Japan. J. Appl. Phys. **32**, L1642–L1644 (1993)
8. S. Nakamura, H. Fujita: Trans. IEE Japan **117** E, 633–636 (1997)
9. M.I. Lutwyche, Y. Wada: Sensors & Actuators A **48**, 127–136 (1995)
10. M.I. Lutwyche, Y. Wada: Appl. Phys. Lett. **48**, 1027 (1995)
11. M.I. Lutwyche, Y. Wada: J. Vac. Sci. Technol. B **13**, 2819–2822 (1995)
12. Y. Wada: Microelectronics J. **29**, 601–611 (1998)
13. N.F. Raley, Y. Sugiyama, T. van Duzer: J. Electrochem. Soc. **131**, 151–171 (1984)
14. J. Itoh, S. Kanemaru, T. Matsukawa: Oyo Buturi **67**, 1390 (1998) [in Japanese]
15. H. Fujii, S. Kanemaru, T. Matsukawa, H. Hiroshima, H. Yokoyama, J. Itoh: Jpn. J. Appl. Phys. **37**, 7182 (1998)
16. G. Hashiguchi, H. Mimura: Jpn. J. Appl. Phys. **33**, L1649–L1650 (1994)
17. G. Hashiguchi, H. Mimura: Jpn. J. Appl. Hys. **34**, 1493 (1995)
18. G. Hashiguchi, M. Goto, M. Mita, H. Toshiyoshi, H. Fujita, D. Kobayashi, J. Endo, Y. Wada: to be summitted to Jpn. J. Appl. Phys.
19. T. Furuhata et al.: *Sub-Micron Gaps Without Sub-Micron Etching*, Technical digest of IEEE, Micro Electro Mechanical Systems Workshop 1991, Nara, Japan (January 1991) pp. 57–62
20. H. Yamada et al.: *Experimental Study of Forces Between a Tunnel Tip and the Graphite Surface*, J. Vac. Sci. Technol. A **6** (2), 293–295 (1988)
21. D.M. Eigler, E.K. Schweizer: Nature **344**, 524–526 (1990)
22. A. Kobayashi, F. Grey, R.S. Williams, M. Aono: Science **259**, 1724–1726 (1993)
23. E.S. Snow, P.M. Campbell, P.J. Mcmarr: Appl. Phys. Lett. **63**, 749–752 (1993)
24. K. Matsumoto: Proc. IEEE **85**, 612–628 (1997)
25. L.L. Sohn, R.L. Willett: Appl. Phys. Lett. **67**, 1552–1555 (1995)
26. W.J. Dressick, J.M. Calvert: Jpn. J. Appl. Phys. **32**, 5829–5839 (1993)

27. M. Ishibashi, S. Heike, H. Kajiyama, Y. Wada, T. Hashizume: Jpn. J. Appl. Phys. **37**, 1565–1569 (1998)
28. Y. Wada, M.I. Lutwyche, M. Ishibashi: SPIE Proc. **2879**, 327–331 (1996)
29. J.W. Lyding, T.-C. Shen, J.S. Hubacek, J.R. Tucker, G.C. Abeln: Appl. Phys. Lett. **64**, 2010–2013 (1994)
30. T. Hashizume, S. Heike, M.I. Lutwyche, S. Watanabe, K. Nakajima, T. Nishi, Y. Wada: Jpn. J. Appl. Phys. **35**, L1085–L1088 (1996)
31. Y. Wada, T. Uda, M.I. Lutwyche, S. Kondo, S. Heike: J. Appl. Phys. **74**, 7321–7328 (1993)
32. Y. Wada: Trans. IEICE 1994, OME 93 **54**, 31–38
33. S. Okazaki: Appl. Surf. Sci. **70/71**, 603–612 (1993)
34. S. Asai, Y. Wada: Proc. IEEE **85**, 505–520 (1998)
35. H. Fujita: Sensors & Actuators A **56**, 105–111 (1996)
36. H. Fujita: Trans. IEE Japan **117** E, 401–406 (1997)
37. S. Heike, T. Hashizume, Y. Wada: Japan. J. Appl. Phys. **34**, L1061–L1063 (1995)
38. H. Ueno: Trans. IEICE E**77**-D, 68–75 (1994)
39. H.J. Mamin, D. Rugar: *Thermomechanical Writing with an Atomic Force Microscope Tip*, Appl. Phys. Lett. **61**, 1003–5 (1992)
40. H.J. Mamin, L.S. Fan, S. Hoen, D. Rugar: *Tip-Based Data Storage Using Micromechanical Cantilevers*, Sensors & Actuators A **48**, 215–9 (1994)
41. B.W. Chui, T.D. Stowe, T.W. Kenny, H.J. Mamin, B.D. Terris, D. Rugar: *Low-Stiffness Silicon Cantilevers for Thermal Writing and Piezoresistive Readback with the Atomic Force Microscope*, Appl. Phys. Lett. **69**, 2767–9 (1996)
42. Q. Ru, J. Endo, T. Tanji, A. Tonomura: *Phase-Shifting Electron Holography by Beam Tilting*, Appl. Phys. Lett. **59**, No. 19, 2372–2374 (1991)

8 Nanoscale Characterization of Nanostructures and Nanodevices by Scanning Probe Microscopy

Takuji Takahashi

Semiconductor nanostructures such as quantum wires and quantum dots are receiving wide attention as new materials for the next generation of semiconductor devices, because many device properties are expected to be improved by quantum confinement effects [1,2]. However, by miniaturizing the geometrical structures down to $\sim$ 10 nm, it becomes more difficult to fabricate such fine structures. It is also harder to characterize them on a real scale of spatial resolution.

Conventional methods for spatial resolution of electrical or optical measurements using electrodes or laser light are limited by electrode size or light wavelength. The size limit of the electrode is around 0.1 μm, even when we use high-performance electron beam lithography techniques. Visible or near infrared light is often used to characterize the optical properties of semiconductors. This is because bandgap energies of typical semiconductors such as Si, Ge, and GaAs are around 1 eV, and this corresponds to a wavelength of order 1 μm. Hence, these methods average the electrical and optical properties over many wires or dots whose typical size is about 10 nm. In contrast, scanning probe microscopy (SPM), which originated from scanning tunneling microscopy (STM) [3] and atomic force microscopy (AFM) [4], has high enough spatial resolution to approach each individual wire or dot structure. In particular, STM and AFM with a conductive tip allow us to study the electrical properties inside nanostructures, and scanning near-field optical microscopy (SNOM) [5–7], which broke the wavelength limit, provides us with a novel way to characterize optical properties.

8.1 Micromachining Technologies in SPM

SPM equipment requires coarse positioning of the sample and tip on a millimeter scale as well as fine positioning on a nanometer scale. This coarse positioning necessitates a wide dynamic range from millimeter to submicrometer as far as movement is concerned and a good level of stability at standstill. In addition, the minimum step in the coarse movement should be smaller than the total range of the fine positioning in the z-direction. For such purposes, a screw movement, a stepping motor, and an inertial driving technique are widely used (see Fig. 8.1 and also Chap. 6). For the fine positioning, piezoelectric materials are mainly used (Fig. 8.2), allowing three-dimensional motion of the tip or the sample with good reproducibility in positioning.

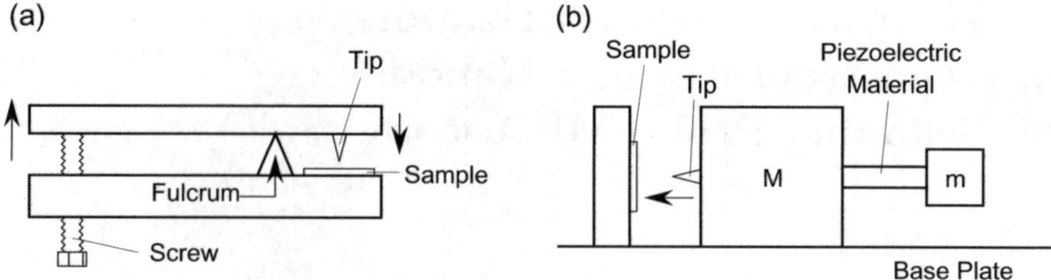

Fig. 8.1. Coarse positioning methods for SPM: (**a**) a screw shaft, which is rotated by a micrometer or a stepping motor, pulling up a tip-mounted plate to make the tip approach the sample, and (**b**) an inertial driving technique, in which two masses are connected by a piezoelectric material

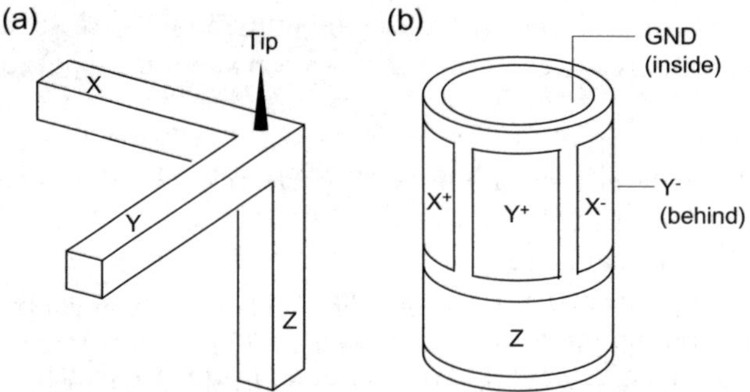

Fig. 8.2. Piezoelectric scanners for SPM: (**a**) a tripod scanner consisting of three bars perpendicular to each other, and (**b**) a tube scanner with some divided electrodes. The voltages applied to the electrodes X^+, X^-, Y^+, and Y^- bend the scanner. The voltages at X^- and Y^- have the opposite signs to those applied at X^+ and Y^+, respectively, and the voltage applied to the Z electrode causes a uniform elongation

A piezoelectric material with high resonant frequency is ideal for the SPM scanner in order to achieve high-speed scanning, whilst avoiding the risk of exciting vibrations. Nowadays, the tube scanner (Fig. 8.2b) is the most popular because of its compact and rigid design, resulting in a high resonant frequency.

The SPM also requires a tip with very fine apex. Mechanical cutting or electrochemical etching (Fig. 8.3) of a metal wire are used to fabricate the single STM tip, and microfabrication techniques are sometimes used to realize twin tips or multiple tip arrays. Moreover, microfabrication techniques are applied to fabricate the AFM tips from semiconductor-based materials like Si, SiO_2, and SiN_x (Fig. 8.4) [8,9], and to fabricate SNOM tips from optical fibers [10]. Micromachining technology is thus one of the key technologies used to develop the SPM, making it possible to apply the latter in a very wide range of fields, as described in the following.

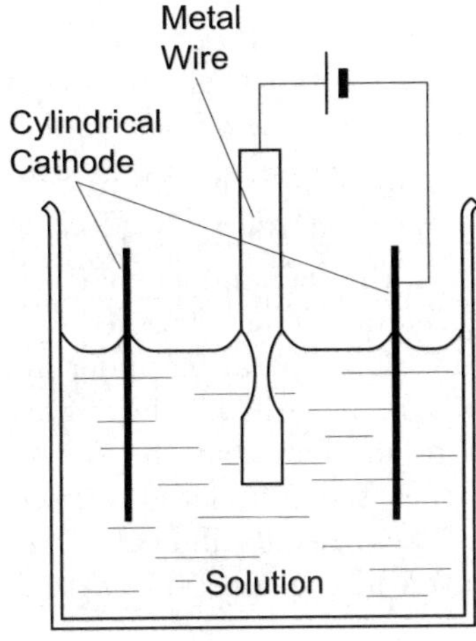

Fig. 8.3. Schematic illustration of the electrochemical etching cell. A positively biased metal wire like tungsten can be etched by a solution of NaOH or KOH

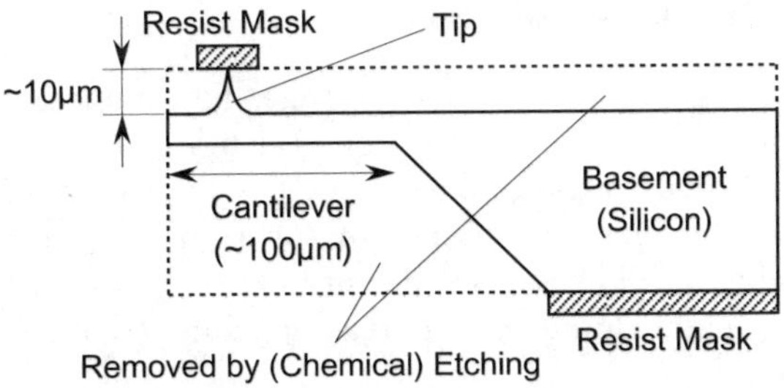

Fig. 8.4. A rough sketch of the microfabrication process for the AFM cantilever and tip. Photolithography and (chemical) etching methods are used to form a cantilever about 100 μm long and a tip about 10 μm high on a base material such as silicon

8.2 Scanning Tunneling Microscopy and Spectroscopy for Semiconductors

Recently, the surface topographies on semiconductors have been widely studied by STM. For instance, the (7×7) reconstruction of the Si (111) surface was observed in the very early stages of STM [11] and this reconstructed surface is often used as a standard sample to calibrate an xy piezo in the STM equipment under ultrahigh vacuum conditions. Since the STM probes the local current on a sample, much research has gone into investigating local electrical properties as well as topography. Such attempts are called scanning tunneling spectroscopy (STS). In this section, we focus on the STM and STS of semiconductor nanostructures such as heterostructures and quantum dot structures.

8.2.1 Topographic Characterization

Topographic images obtained by STM in the constant current mode are affected by changes in local conductance. Tanaka et al. reported cross-sectional imaging of the GaInAs/InP and AlAs/GaAs multi-quantum wells (MQW) by STM in air, observing topographic corrugations. They attribute this corrugation to the tunnel conductance difference between the well and the barrier, resulting from their valence band offset [12]. Salemink et al. observed the cross-section of AlGaAs/GaAs heterostructures by STM operating under ultrahigh vacuum conditions. In the filled state image, the AlGaAs layer was imaged lower than the GaAs layer. This is due to the fact that the Al–As bond is more ionic than the Ga–As bond and that the As-related surface filled states are altered near the Al atoms [13]. These results indicate that the composition distribution can be visualized even in the constant current mode of STM.

8.2.2 Scanning Tunneling Spectroscopy (STS)

The tunneling current in STM depends on the bias voltage. For tunneling between two metal electrodes in the low-bias limit (of order 1 mV), the tunneling current is linearly proportional to the applied bias. For high bias (of order 1 V), on the other hand, the bias-dependence of the tunneling current does not exhibit Ohmic behavior. The surface density of states and the lowering of the effective barrier height dominate the bias-dependence of the tunneling current.

Figure 8.5 shows schematic energy diagrams for a sample and tip separated by a narrow tunneling (vacuum) gap. For zero bias (Fig. 8.5a) the Fermi levels of the tip and sample are equal at equilibrium. When a positive bias is applied to the sample (Fig. 8.5b), electron tunneling from the tip to the sample probes unoccupied states in the local density of states (LDOS) of the sample, represented by a wavy line in Fig. 8.5. Of course, the LDOS of the tip influences the tunneling current, but in the case of a metal tip and semiconductor sample, the LDOS of the sample dominates the tunneling current property, because the LDOS of the tip is much larger than that of the sample. Under negative sample bias conditions, the occupied states in the sample LDOS can be probed (Fig. 8.5c). As a result, the sample LDOS is scanned by varying the bias voltage. Since the total tunneling current is given by integrating over all tunneling electrons, the first derivative $\mathrm{d}I/\mathrm{d}V$ of the tunneling current I with respect to the bias V reflects the LDOS at that bias condition [14]. Feenstra et al. showed that the normalized differential conductance $\mathrm{d}I/\mathrm{d}V/(I/V)$ compensates for both the bias-dependence of the tunneling probability and the change in the tip–sample separation, and that this method is more effective for estimating the existence of a bandgap and energy states within the bandgap [15].

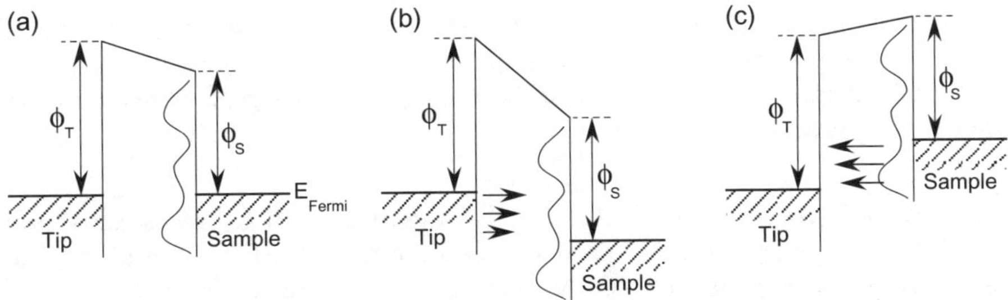

Fig. 8.5. Schematic energy diagrams for tip and sample: (**a**) tip and sample at equilibrium, separated by a narrow tunneling gap, (**b**) positive sample bias where electrons tunnel from the tip to the sample, and (**c**) negative sample bias where electrons tunnel from the sample to the tip

STS on GaAs/AlGaAs Heterostructures. Cross-sectional STM was performed on *p*–*n* multilayers of GaAs [16,17], in which each layer was identified by the conductance difference between the *n*-type layer and the *p*-type layer (Fig. 8.6), and on GaAs/AlGaAs multilayers [17,18], where the difference in the conductance value or the difference in the bandgap in the tunnel-

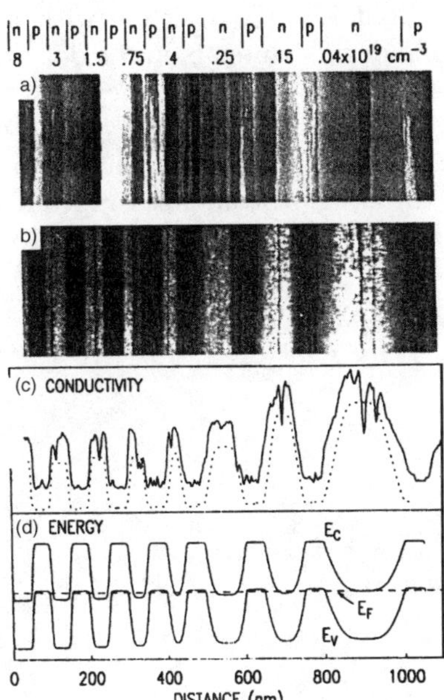

Fig. 8.6. The GaAs doping supperlattice with *n*-type doping concentrations listed, and a *p*-type concentration of 3×10^{19} cm^{-3}. The constant-current STM topography (**a**) and conductivity image (**b**) were acquired with a sample voltage of -2.0 V. Line profiles of the measured conductivity (*solid line*) and model calculation (*dotted line*) are shown in (**c**), and the computed band diagram is shown in (**d**) [13]

ing spectra was clearly observed. These results indicate that layer identification is possible through STS measurements, although they have a disadvantage in that real-time identification is difficult, because STS measurements take a long time.

STS on Self-Assembled InAs Dot Structures. When InAs is grown on a (001) GaAs substrate by molecular beam epitaxy (MBE) or metal-organic chemical vapor deposition (MOCVD), InAs dot structures self-assemble as a result of strain effects due to the fact that the lattice constant of InAs is about 7% larger than that of GaAs. The typical diameter and height of the dots are 20–30 nm and 10 nm, respectively, in which the quantum confinement effect of electrons or holes will appear. Therefore, the self-assembled formation of the InAs dot is one of the best approaches to nanofabrication of quantum dot structures.

The SPM allows us to approach each dot and characterize its electrical properties. Legrand et al. performed STM and STS on cross-sections of stacked self-assembled InAs dots. They showed that the bandgap appearing in the $\mathrm{d}I/\mathrm{d}V/(I/V)$ curves on the InAs dot is narrower than that occurring far from the dots (Fig. 8.7) [19]. Each dot has a lens-like shape, with width and height 26 nm and 4 nm, respectively. According to the STS measurement, the bandgap far from the dot was evaluated at about 1.4 eV, consistent with the bandgap value of bulk GaAs. On the other hand, the bandgap on the dot was about 1.25 eV, much larger than the bulk InAs value (≈ 0.4 eV). The bandgap value on the dot includes the bulk bandgap and the quantum energy levels in both the conduction band and the valence band. By decreasing the dot size below 100 nm, the quantum confinement of the electrons and holes appears in

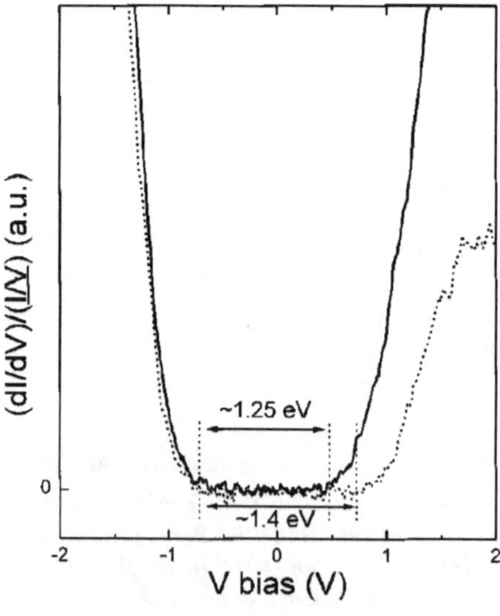

Fig. 8.7. Local spectroscopy measured on an InAs quantum dot (*solid line*) and on GaAs far from the dots (*dotted line*) [16]

a remarkable way, leading to a widening of the overall gap. The bulk bandgap in InAs is widened by the strain effect owing to the lattice mismatch between InAs and GaAs, and this also enhances the total bandgap. Atomic resolution STM images confirmed this strain effect in the InAs dot region.

Ballistic Electron Emission Microscopy (BEEM) on Quantum Dots. Ballistic electron emission microscopy (BEEM) is an application of STM. For BEEM measurements, a base contact and a collector contact are formed on the top and back surfaces of a sample, and an STM tip acts as an electron emitter (Fig. 8.8). The STM tip height is controlled by the tunneling current flowing between the STM tip and the base contact. Some of the emitted electrons from the STM tip flow ballistically towards the back contact if the electron energy exceeds the Schottky barrier height at the metal–semiconductor interface. Thus the BEEM current property can be used to characterize the potential profiles inside the sample.

Figure 8.9a shows a schematic cross-sectional view of a single InAs self-assembled quantum dot buried beneath an Au/GaAs interface, whilst Fig. 8.9b shows the BEEM current spectra obtained on and off the dot [20]. In this sample, quantized energy levels probably exist in the dot structures. When the tip bias is aligned with such a quantized energy level, electrons flow resonantly from the STM tip to the collector contact through the quantized

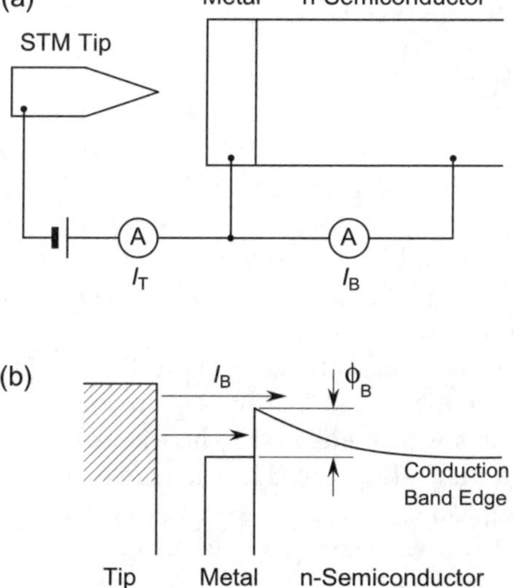

Fig. 8.8. Schematic view (**a**) and energy diagram (**b**) of the BEEM. Electrons tunneling from the tip into the thin metal film are detected as the tip current I_{T}, and I_{T} is used to control the tip height, as in conventional STM. Those electrons which reach the metal–semiconductor interface ballistically may cross it to form the BEEM current I_{B} if the energy exceeds the Schottky barrier height ϕ_{B}

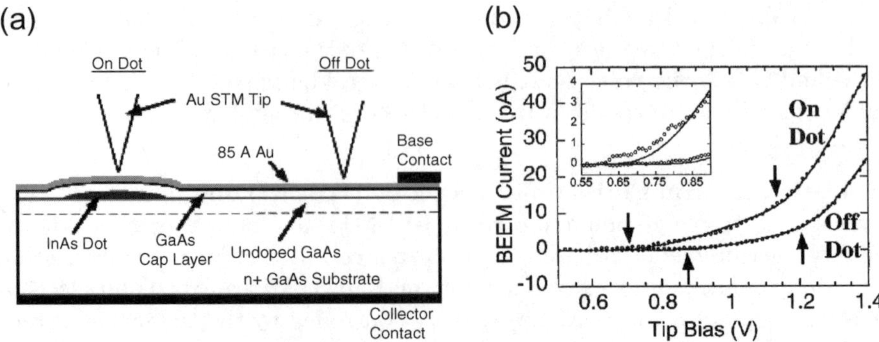

Fig. 8.9. (**a**) Schematic cross-sectional view of the sample structure and (**b**) the BEEM spectra obtained on and off the dot [17]

energy level, even if the electron energy is below the Schottky barrier height. This configuration is very similar to that in the resonant tunneling diode. As a result, a threshold shift and some distinctive features are observed in the BEEM current on the dot, at energies below the Schottky barrier threshold, as shown in Fig. 8.9b. In contrast to this situation, AlInP quantum dots buried in GaP behave like barriers for the BEEM current, because the bandgap of AlInP is wider than that of GaP [21].

8.2.3 STM Luminescence from Nanostructures

There have been several attempts to treat the STM tip as an electron/hole emitter and to detect electroluminescence from nanostructures excited by tunneling electrons/holes. Murashita developed a conductive transparent tip to collect luminescence at low temperature using an optical fiber, and succeeded in detecting tunneling luminescence from a cross-section of GaAs/AlAs MQW structures, as well as STM topography at 10 K (see Fig. 8.10) [22]. The luminescence from the GaAs quantum wells induced by the STM current was clearly observed, as can be seen in the figure.

Tunneling luminescence from the InAs self-assembled dots on (001) GaAs was obtained under ultrahigh vacuum conditions at 120 K [23]. In this case, the STM tip was used as a hole injector and the emitted light was collected by a lens and led by an optical fiber to a photodetector. The estimated spatial resolution of this system is about 40 nm, and luminescence was successfully obtained from a single quantum dot with a very sharp single peak.

8.2.4 Combination of STM/STS and Laser Illumination

One can investigate optical properties by laser light irradiation of the tunneling gap between a sample and the STM tip. Several mechanisms may

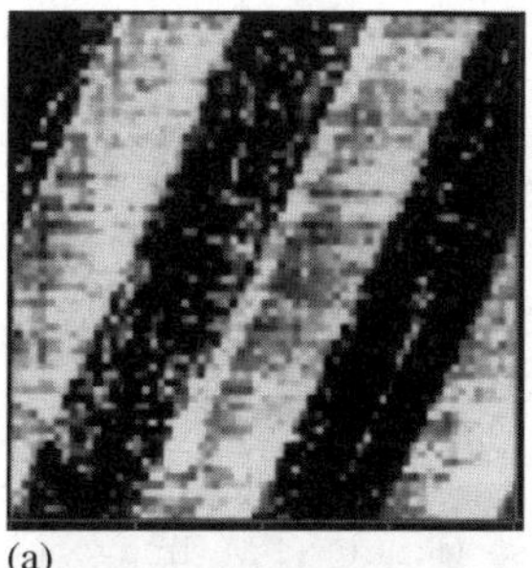

(a)

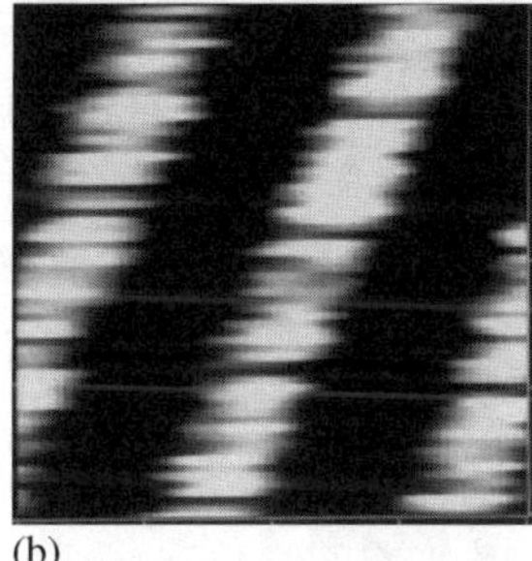

(b)

Fig. 8.10. (**a**) Topographic and (**b**) tunneling luminescence images of a cross-section of GaAs/AlAs multiple quantum well structures taken simultaneously at 10 K by STM with a conductive transparent tip [22]

influence the tunneling current, e.g., thermal expansion of the tip and sample, the generation of surface plasmons, surface photovoltaic effects, and so on. Especially in semiconductors, incident laser irradiation with energy at or above the bandgap creates electron–hole pairs within the absorption depth. The created electrons and holes reduce the resistance in the semiconductor sample, and this allows us to obtain STM images even on semi-insulating semiconductor samples [24]. For undoped GaAs/AlGaAs multilayers, selective excitation of the GaAs region, whereby the resistance in the GaAs layer is selectively reduced, is possible if an appropriate laser wavelength is chosen so as to be absorbed only in the GaAs region. Consequently, by cross-sectional STM imaging of GaAs/AlGaAs heterostructures under laser irradiation, the height contrast between the GaAs layer and the AlGaAs layer have been clearly obtained in the constant-current mode of operation (Fig. 8.11) [25]. Such a contrast disappears when the laser irradiation is absent. This result implies that selective excitation by laser light can visualize compositional differences on the nanometer scale.

If a semiconductor has a surface depletion layer, as in the case of Si and GaAs, owing to surface Fermi-level pinning in the bandgap, photoexcited electron–hole pairs in the depletion layer become separated by the built-in electric field. This results in a surface photovoltaic effect [26]. The surface photovoltaic effect on GaAs is relatively large compared with InAs surfaces, where an opposing electron accumulation layer exists [27]. In this sense, the GaAs (001) surface covered by InAs self-assembled dot structures is very interesting. On this surface, GaAs, which tends to be depleted, is covered

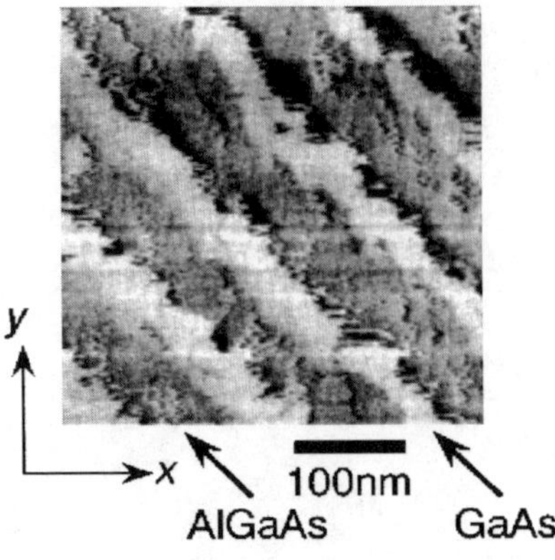

Fig. 8.11. Cross-sectional STM image of undoped GaAs/AlGaAs heterostructures obtained under laser irradiation at wavelength 800 nm and power 145 mW [25]

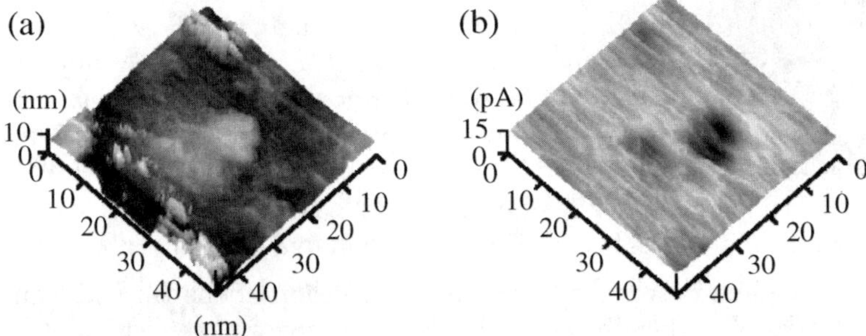

Fig. 8.12. (**a**) Topographic and (**b**) photoinduced current images on the InAs-dot-covered (001) GaAs surface [29]

by InAs, which accumulates electrons near the surface, with nanometer scale lateral distribution. Takahashi et al. investigated such a surface by STS under laser irradiation. They concluded that surface depletion under the InAs dot is released and that the surface Fermi level shifts towards the conduction band edge of GaAs, while the surface under the wetting layer, far from the dot, is depleted just as in bulk GaAs [28]. On a similar sample, a photoinduced current image was successfully obtained simultaneously with a topographic image during laser irradiation modulated by an optical chopper (Fig. 8.12). This indicates that surface depletion beneath the InAs dot was reduced [29]. This result is perfectly consistent with the previous one.

8.3 Atomic Force Microscopy (AFM) on Semiconductor Nanostructures

Nowadays, atomic force microscopy (AFM) [4] is more widely used than STM, because AFM can observe any kind of sample regardless of whether it is a conductor or an insulator, whereas the sample must be conductive for STM measurements. A lot of work has been reported on topographic characterization of semiconductor nanostructures relating to the dot formation process, selective area growth and etching, and other topics. In addition, electrical properties and topography can be characterized simultaneously by using a conductive AFM tip. The latter is generally made of heavily doped Si or metal coating on a conventional AFM tip. In this section, we will review some AFM activities with a conductive tip, such as a current probe, a capacitance probe, or an electrostatic force detector, on semiconductor nanostructures.

8.3.1 AFM with a Conductive Tip as a Current Probe

If a bias voltage is applied between a sample and a conductive AFM tip in the contact mode, the AFM tip acts as a local current probe. When the metal-coated tip touches a semiconductor surface, it can be treated as a nanometer scale Schottky contact. Tanaka et al. used AFM with a conductive tip to investigate the I–V characteristics on self-assembled InAs quantum dot structures grown on (001) GaAs. They showed that the conductivity increased with increasing dot size (Fig. 8.13) [30]. This result indicates that the InAs dot on the surface acts as a small electrode and modulates the barrier height at the interface between InAs and GaAs.

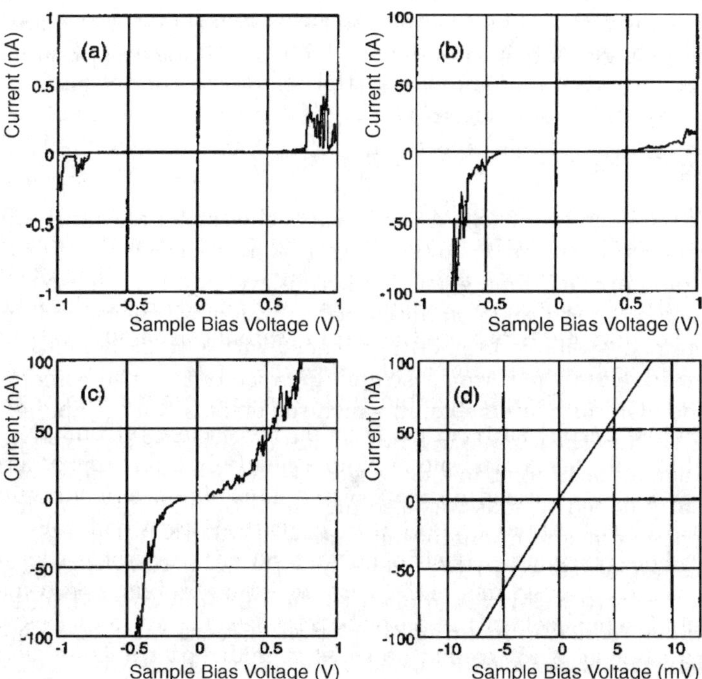

Fig. 8.13. Measured I–V characteristics on (**a**) a wetting layer, (**b**) a 20 nm dot, (**c**) a 50 nm dot, and (**d**) a 100 nm dot [30]

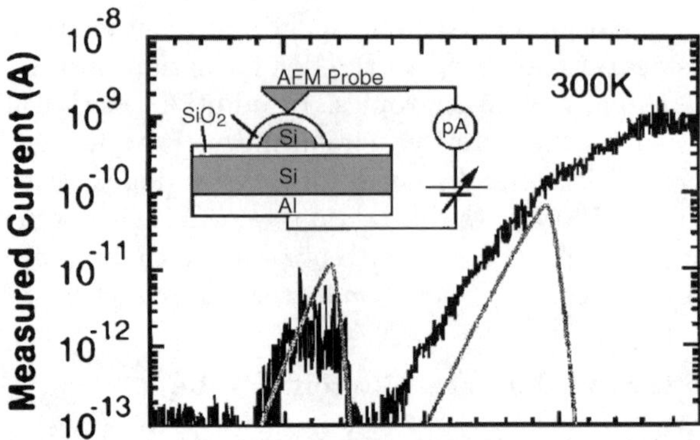

Fig. 8.14. Measured *I–V* characteristics on Si/SiO_2 double-barrier structures with an Si dot of height 2.7 nm. The *dotted curve* is the calculated characteristic based on the device structures [31]

On the other hand, *I–V* characterization was also performed on Si quantum dot structures spontaneously fabricated on an SiO_2 surface [31]. Here, a resonant tunneling diode structure with double barriers was formed and a current resonance was clearly observed in the current flowing between the conductive AFM tip and the Si substrate (Fig. 8.14). The observed current bump and negative conductance can be interpreted in terms of the resonant tunneling through a single Si quantum dot in the SiO_2/Si-dot/SiO_2 double barrier structures.

8.3.2 Scanning Capacitance Microscopy (SCM)

Scanning capacitance microscopy (SCM) [32], which consists of a conductive AFM tip and a capacitance sensor, allows us to study near-surface band profiles originating from impurity doping, surface Fermi-level pinning, and so on. The two-dimensional dopant profile on implanted Si wafers was measured by SCM and the obtained profile is in good agreement with lateral simulation profiles [33]. Edwards et al. performed scanning capacitance spectroscopy (SCS) based on the SCM measurements, in which the dc bias voltage applied between the SCM tip and the sample was cycled and $\mathrm{d}C/\mathrm{d}V$ was obtained as a function of voltage at each point in the image [34]. An n-type Si substrate which had differently doped regions was also measured by SCM [35]. In this SCM image, n-type regions additionally implanted with both P^+ and As^+, and p-type regions implanted with BF^{2+} can be clearly recognized, as shown in Fig. 8.15. Very recently, SCM was applied to the InAs-dot-covered (001) GaAs surface and the reduction in surface depletion under the dot was confirmed by the enhancement of the capacitance value when the tip was located just on the dot [36].

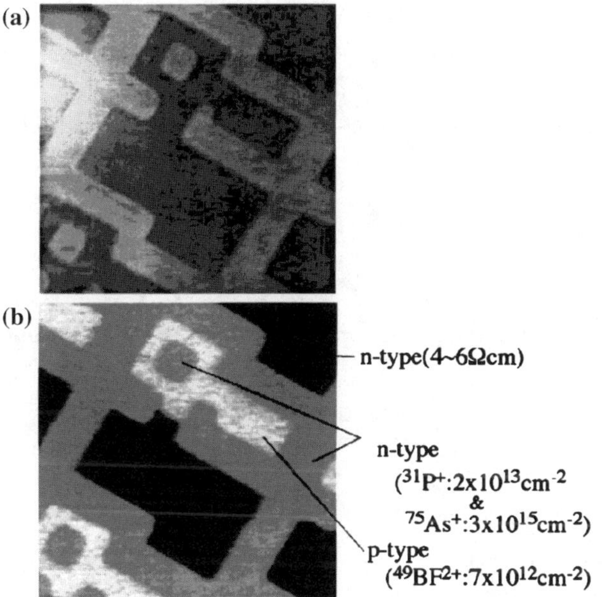

Fig. 8.15. (**a**) AFM image and (**b**) SCM image obtained simultaneously on a sample with different dopant types and densities [35]

8.3.3 Electrostatic Force Detection

In contact mode AFM, the van der Waals force operates between the AFM tip and the sample, and when a bias voltage is applied between them, there is also an electrostatic force. If an ac bias is applied, the electrostatic force component can be extracted from the total working force by a lock-in detection technique. In addition, the bias polarity dependence of the electrostatic force informs us about the near-surface band profiles, such as the degree of surface depletion [37]. Figure 8.16 schematically illustrates the polarity dependence of the electrostatic attractive force F_{AC} on the sample bias voltage. In general, F_{AC} is modulated at twice the frequency of the applied ac bias, as shown in Fig. 8.16a. On the other hand, when we assume an n-type and surface-depleted sample, the gap width d between the tip and the conductive region in the sample depends on the polarity of the bias, as illustrated at the bottom of Fig. 8.16b. F_{AC} is then modulated as indicated by the solid line, in which an f-frequency component appears, as indicated by the broken line.

Figure 8.17a shows the topography and Figs. 8.17b and c the electrostatic force images at frequencies of $2f$ and f, respectively, on the InAs-dot-covered (001) GaAs surface. The $2f$ and f components on the dot are stronger and weaker than those outside the dot, respectively. The latter means that the gap width d is almost independent of the polarity of the bias, while the former means that the working electrostatic force is itself enhanced in the

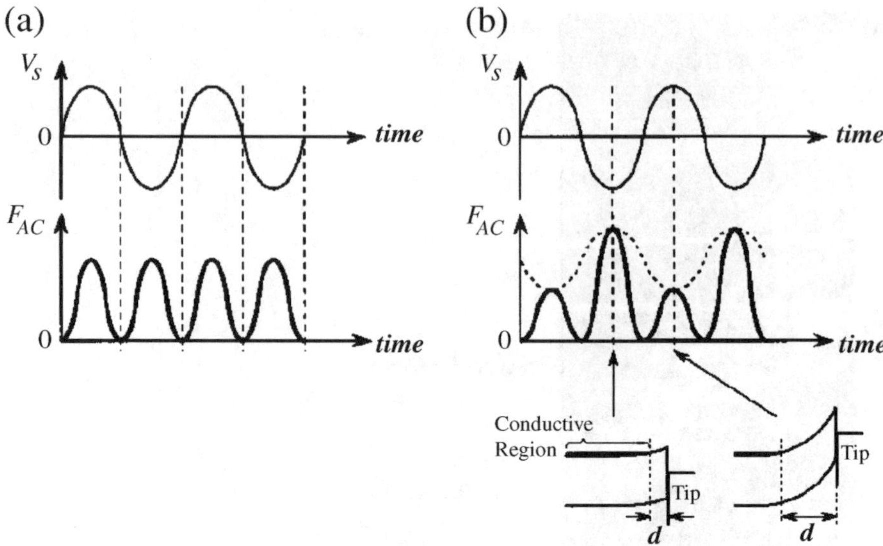

Fig. 8.16. Schematic illustration of the polarity dependence of the electrostatic force F_{AC} on the sample bias voltage V_{S} in (**a**) the general case and (**b**) an n-type sample with surface depletion

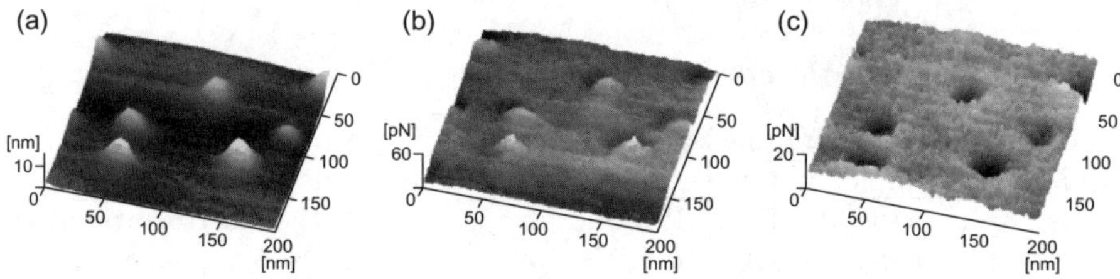

Fig. 8.17. (**a**) Topographic image of the InAs dots on (001) GaAs and electrostatic force images obtained at frequencies (**b**) $2f$ and (**c**) f [37]

dot-covered area. This result implies a reduction in the average gap distance d beneath the dot, that is, a reduction in the surface depletion thickness. It is in good agreement with previous STM/STS studies [28,29].

8.3.4 Kelvin Probe Force Microscopy (KFM)

Kelvin probe force microscopy (KFM) is another application of electrostatic force measurements on AFM [38]. In KFM, a dc offset bias is additionally applied between the tip and the sample and adjusted to nullify the f component of the electrostatic force. As a result, the dc bias corresponds to the contact potential difference (CPD) between the tip and the sample and a CPD image can be obtained simultaneously with a topography.

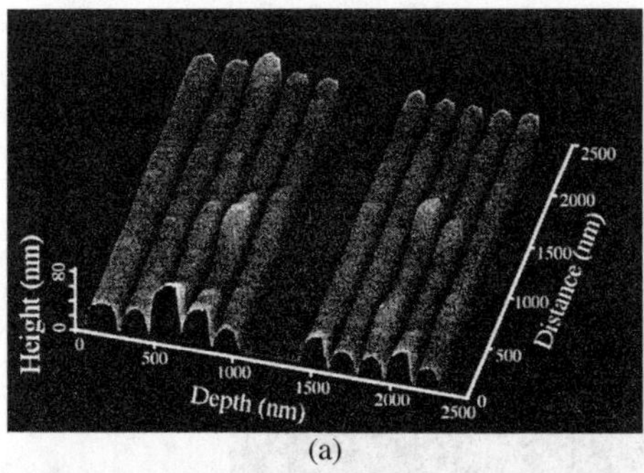

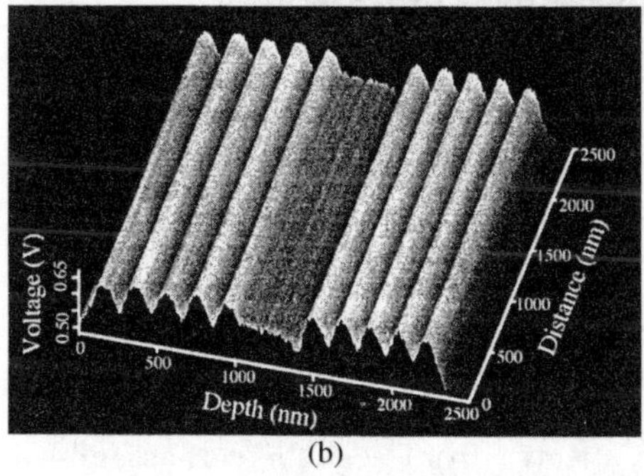

Fig. 8.18. (**a**) Topographic image and (**b**) the corresponding potential image of GaAs/AlAs MQW structures [41]

A cross-section of the Si *p*–*n* junction [39], GaAs/AlGaAs *n*–*i*–*p*–*i* MQW structures [40], and undoped GaAs/AlAs or InAlAs/InGaAs heterostructures [41] were characterized by KFM, and the surface potential distributions originating from the band diagrams were successfully observed (Fig. 8.18). The surface potential distribution was clearly observed from measurements on the corrugated InAs surface grown on (110) GaAs substrates, implying a carrier concentration along the edges of the corrugation steps [42].

On the other hand, the potential profiles in semiconductor devices in operation were investigated by KFM. Two-dimensional potential images were obtained of the cleaved HEMT (high electron mobility transistor) [43] and LED (light emitting diode) [44].

8.4 Scanning Near-field Optical Microscopy (SNOM)

Scanning near-field optical microscopy (SNOM) [5–8] provides a novel way to characterize optical properties whilst breaking the wavelength limit in resolution. A tapered micropipette or optical fiber is widely used as a SNOM tip. Developments in the shear force operation mode [45] have particularly contributed to expanding the field of application of SNOM.

Toda et al. performed SNOM measurements on GaAs quantum dot structures embedded in AlGaAs barriers fabricated by selective epitaxial growth

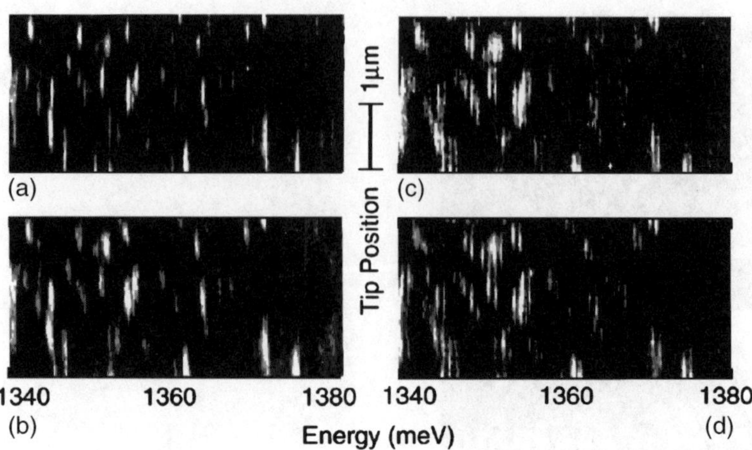

Fig. 8.19. Spatial evaluation of the unpolarized near-field PL spectra at various magnetic fields: (**a**) 0 T, (**b**) 4 T, (**c**) 8 T, and (**d**) 10 T [47]

on SiO_2-patterned GaAs (100) substrates [46]. As a result, spatially resolved photoluminescence (PL) spectra and spectrally resolved PL images were successfully obtained.

The same group measured the PL spectra from a single self-assembled InAs quantum dot by SNOM at low temperature under the application of high magnetic fields [47,48]. The Zeeman splittings due to the spin polarities of carriers in a single InAs dot were clearly observed (Fig. 8.19), whilst circularly-polarized PL resolved carriers with different spin polarities.

8.5 Nanofabrication Processes Using STM/AFM

SPM techniques are also used in many different ways for nanofabrication processes. Eigler et al. demonstrated STM manipulation of Xe atoms adsorbed on Ni (110) surfaces to create some letters of the alphabet (Fig. 8.20) [49]. In this experiment, the STM tip was scanned farther from the sample surface during topography observation in order to avoid unintentional displacement of atoms, while the STM tip was brought very close to the sample during atom manipulation, as shown in Fig. 8.21. Using a similar manipulation process, Crommie et al. succeeded in producing a circular disposition of Fe atoms on a Cu (111) surface, known as a quantum corral (Fig. 8.22), where some ripples in the surface local density of states of electrons were clearly observed [50].

Marrian et al. patterned a polydiacetylene negative electron beam resist by electron exposure from an STM tip in vacuum at pressures in the 10^{-8} torr range [51]. They succeeded in fabricating a minimum feature size of about 20 nm by choosing an appropriate electron dose condition.

When a high bias is applied between the STM tip and a sample, the field evaporation of an atom occurs from the sample or tip surface. Hosoki et al.

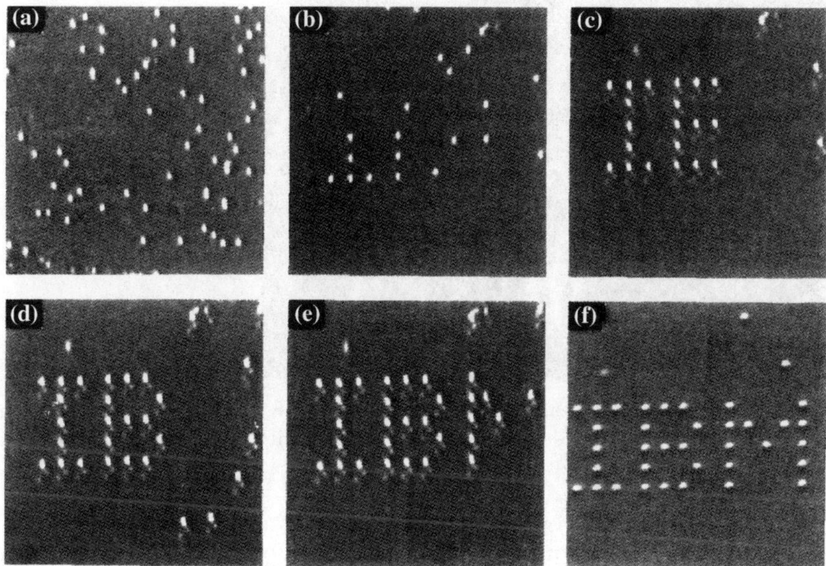

Fig. 8.20. Sequence of STM images taken during the build-up of a patterned array of the letters I, B, and M constructed from Xe atoms adsorbed on the Ni (110) surface. Each letter measures 5 nm from top to bottom [49]

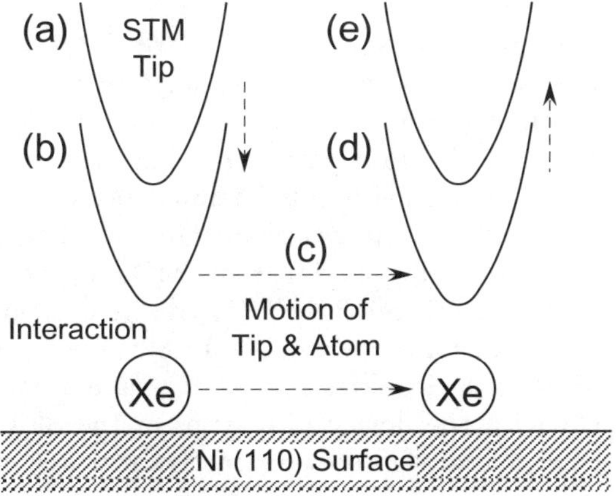

Fig. 8.21. Schematic illustration of the manipulation process for sliding an atom across a metal surface using an STM tip. (**a**) The tip is placed far from the sample surface in order to observe the topography before manipulation. (**b**) The tip is brought closer to the sample and the attractive tip–atom interaction comes into play. (**c**) The atom is manipulated by the tip movement to the desired destination (**d**). (**e**) the tip is retracted, leaving the atom at a new location

Fig. 8.22. A so-called quantum corral, consisting of 48 Fe atoms adsorbed on a Cu (111) surface. Some ripples in the local surface density of states of electron are clearly visible inside the corral [50]

reported the field evaporation of S atoms from an MoS_2 crystal by STM, the field evaporation of Au atoms from an Au-coated AFM cantilever to SiO_2/Si surfaces, and the phase change of GeSbTe from amorphous to crystalline by a local heating method in SNOM [52,53]. In the first method, fully atomic-scale nanofabrication was achieved, while the resolution in the other two methods was limited to the order of nanometers or several tens of nanometers. Mamin et al. also demonstrated the field evaporation of Au atoms from the Au STM tip to form complex patterns [54].

On the other hand, local oxidation of the sample surface by SPM via anode oxidation processes promises wide applicability to nanofabrication. Matsumoto et al. applied STM to form very fine oxidized Ti lines by the tip-induced local oxidation process and succeeded in fabricating a single electron transistor [55]. In their process, they prepared a very thin Ti film deposited on an SiO_2 surface, and a negatively biased STM tip was scanned on it. Then a TiO_x fine line was formed by tip-induced oxidation. Since this TiO_x line acts as a tunneling barrier, a small Ti island surrounded by TiO_x can be formed, in which the Coulomb blockade phenomenon appears. Figure 8.23 shows a schematic illustration of the fabricated device structure acting as a single electron transistor (SET), together with the device properties of the SET, indicating a clear Coulomb staircase, one of the most distinctive features of the SET.

A similar oxidation process can also be applied to semiconductor heterostructures. Direct STM oxidation on Si surfaces and pattern transfer have been demonstrated by Dagata et al. [56]. This method was applied to GaAs surfaces by Snow et al. [57]. Direct oxidation by AFM has also been demonstrated on GaAs [58] and InAs/AlGaSb [59] surfaces.

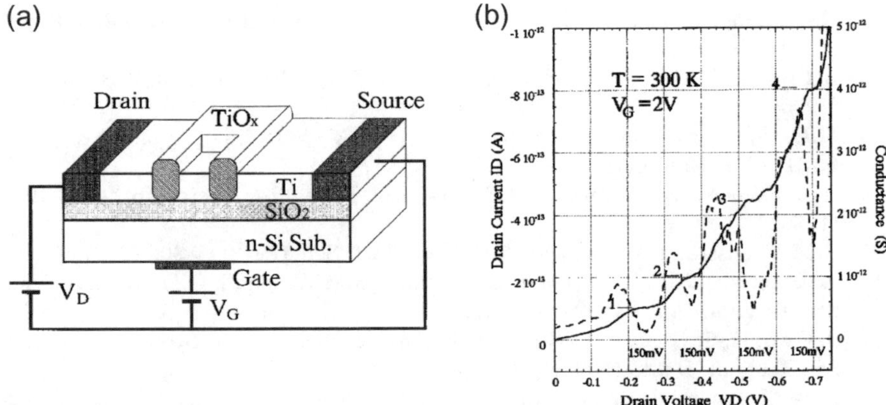

Fig. 8.23. (**a**) Schematic illustration of a single electron transistor fabricated by the STM nano-oxidation process. (**b**) Drain current–voltage characteristics (*solid line*) and conductance (*dashed line*) of the SET at room temperature [55]

8.6 Concluding Remarks

The proposal of quantum wires and dots and the invention of STM were independent, but occurred simultaneously in the early 1980s. This coincidence may well have accelerated their progress in recent years. SPM technologies have made it possible to characterize quantum nanostructures on a real scale. Novel ways of fabricating quantum nanostructures have been developed on the basis of this characterization, and the demands for new methods to characterize certain other physical properties have led to yet more developments in SPM technology. This spiralling relationship holds the door wide open for nanometer scale science and technology in the 21st century.

References

1. H. Sakaki: Jpn. J. Appl. Phys. **19**, L735–L738 (1980)
2. Y. Arakawa, H. Sakaki: Appl. Phys. Lett. **40**, 939–941 (1982)
3. G. Binning, H. Rohrer, Ch. Gerber, E. Weibel: Phys. Rev. Lett. **49**, 57–61 (1982)
4. G. Binning, C.F. Quate: Phys. Rev. Lett. **56**, 930–933 (1986)
5. D.W. Pohl, W. Denk, M. Lanz: Appl. Phys. Lett. **44**, 651–653 (1984)
6. U. Durig, D.W. Pohl, F. Rohner: J. Appl. Phys. **59**, 3318–3327 (1986)
7. E. Betzig, M. Isaacson, A. Lewis: Appl. Phys. Lett. **51**, 2088–2090 (1987)
8. T.R. Albrecht, S. Akamine, T.E. Carver, C.F. Quate: J. Vac. Sci. Technol. A **8**, 3386–3396 (1990)
9. O. Wolter, Th. Bayer, J. Greschner: J. Vac. Sci. Technol. B **9**, 1353–1357 (1991)
10. T. Pangaribuan, K. Yamada, S. Jiang, H. Ohsawa, M. Ohtsu: Jpn. J. Appl. Phys. **31**, L1302–L1304 (1992)

11. G. Binning, H. Rohrer, Ch. Gerber, E. Weibel: Phys. Rev. Lett. **50**, 120–123 (1983)
12. I. Tanaka, T. Kato, S. Ohkouchi, F. Osaka: J. Vac. Sci. Technol. A **8**, 567–570 (1990)
13. M.B. Johnson, U. Maier, H.-P. Meier, H.W.M. Salemink: Appl. Phys. Lett. **63**, 1273–1275 (1993)
14. R.J. Hamers, R.M. Tromp, J.E. Demuth: Phys. Rev. Lett. **56**, 1972–1975 (1986)
15. R.M. Feenstra, J.A. Stroscio, A.P. Fein: Surf. Sci. **181**, 295–306 (1987)
16. A. Vaterlaus, R.M. Feenstra, P.D. Kirchner, J.M. Woodall, G.D. Pettit: J. Vac. Sci. Technol. B **11**, 1502–1508 (1993)
17. S. Gwo, K.-J. Chao, A.R. Smith, C.K. Shih, K. Sadra, B.G. Streetman: J. Vac. Sci. Technol. B **11**, 1509–1513 (1993)
18. H.W.M. Salemink, M.B. Johnson, O. Albrektsen: J. Vac. Sci. Technol. B **12**, 362–368 (1994)
19. B. Legrand, B. Grandidier, J.P. Nys, D. Stievenard, J.M. Gerard, V. Thierry-Mieg: Appl. Phys. Lett. **73**, 96–98 (1998)
20. M.E. Rubin, G. Medeiro-Ribeiro, J.J. O'Shea, M.A. Chin, E.Y. Lee, P.M. Petroff, V. Narayanamurti: Phys. Rev. Lett. **77**, 5268–5271 (1996)
21. C.V. Reddy, V. Narayanamurti, J.H. Ryou, U. Chowdhury, R.D. Dupuis: Appl. Phys. Lett. **76**, 1437–1439 (2000)
22. T. Murashita: J. Vac. Sci. Technol. B **15**, 32–37 (1997)
23. K. Yamanaka, K. Suzuki, S. Ishida, Y. Arakawa: Appl. Phys. Lett. **73**, 1460–1462 (1998)
24. G.F.A. van de Walle, H. van Kempen, P. Wyder, P. Davidsson: Appl. Phys. Lett. **50**, 22–24 (1987)
25. T. Takahashi, M. Yoshita, H. Sakaki: Appl. Phys. Lett. **68**, 502–504 (1996)
26. R.J. Hamers, K. Markert: Phys. Rev. Lett. **64**, 1051–1054 (1990)
27. T. Takahashi, M. Yoshita: Appl. Phys. Lett. **68**, 3479–3481 (1996)
28. T. Takahashi, I. Kamiya, H. Sakaki: Appl. Phys. A **66**, S1055–S1058 (1998)
29. H. Yamamoto, I. Kamiya, T. Takahashi: Jpn. J. Appl. Phys. **38**, 3871–3874 (1999)
30. I. Tanaka, I. Kamiya, H. Sakaki, N. Qureshi, S.J. Allen Jr., P.M. Petroff: Appl. Phys. Lett. **74**, 844–846 (1999)
31. M. Fukuda, K. Nakagawa, S. Miyazaki, M. Hirose: Appl. Phys. Lett. **70**, 2291–2293 (1997)
32. J.R. Matey, J. Blanc: J. Appl. Phys. **57**, 1437–1444 (1985)
33. Y. Huang, C.C. Williams, J. Slinkman: Appl. Phys. Lett. **66**, 344–346 (1995)
34. H. Edwards, R. McGlothlin, R.S. Martin, E.U.M. Gribelyuk, R. Mahaffy, C.K. Shih, R.S. List, V.A. Ukraintsev: Appl. Phys. Lett. **72**, 698–700 (1998)
35. T. Yamamoto, Y. Suzuki, M. Miyashita, H. Sugimura, N. Nakagiri: J. Vac. Sci. Technol. B **15**, 1547–1550 (1997)
36. H. Yamamoto, T. Takahashi, I. Kamiya: to be published in Appl. Phys. Lett.
37. T. Takahashi, T. Kawamukai, I. Kamiya: Appl. Phys. Lett. **75**, 510–512 (1999)
38. M. Nonnenmacher, M.P. O'Boyle, H.K. Wickramashinghe: Appl. Phys. Lett. **58**, 2921–2123 (1991)
39. A. Kikukawa, S. Hosaka, R. Imura: Appl. Phys. Lett. **66**, 3510–3512 (1995)
40. A. Chavez-Pirson, O. Vatel, M. Tanimoto, H. Ando, H. Iwamura, H. Kanbe: Appl. Phys. Lett. **67**, 3069–3071 (1995)

41. T. Usunami, M. Arakawa, S. Kishimoto, T. Mizutani, T. Kagawa: Jpn. J. Appl. Phys. **37**, 1522–1526 (1998)
42. T. Takahashi, T. Kawamukai, S. Ono, T. Noda, H. Sakaki: Jpn. J. Appl. Phys. **39**, 3721–3723 (2000)
43. M. Arakawa, S. Kishimoto, T. Mizutani: Jpn. J. Appl. Phys. **36**, 1826–1829 (1997)
44. R. Shikler, T. Meoded, N. Fried, Y. Rosenwaks: Appl. Phys. Lett. **74**, 2972–2974 (1999)
45. E. Betzig, P.L. Finn, J.S. Weiner: Appl. Phys. Lett. **60**, 2484–2486 (1992)
46. Y. Toda, M. Kourogi, M. Ohtsu, Y. Nagamune, Y. Arakawa: Appl. Phys. Lett. **69**, 827–829 (1996)
47. Y. Toda, S. Shinomori, K. Suzuki, Y. Arakawa: Appl. Phys. Lett. **73**, 517–519 (1998)
48. Y. Toda, S. Shinomori, K. Suzuki, Y. Arakawa: Phys. Rev. B **58**, R10147–R10150 (1998)
49. D.M. Eigler, E.K. Schweizer: Nature **344**, 524–526 (1990)
50. M.F. Crommie, C.P. Lutz, D.M. Eigler: Science **262**, 218–220 (1993)
51. C.R.K. Marrian, R.J. Colton: Appl. Phys. Lett. **56**, 755–757 (1990)
52. S. Hosoki, S. Hosaka, T. Hasegawa: Appl. Surf. Sci. **60/61**, 643–647 (1992)
53. S. Hosaka, S. Hosoki, T. Hasegawa, H. Koyanagi, T. Shintani, M. Miyamoto: J. Vac. Sci. Technol. B **13**, 2813–2818 (1995)
54. H.J. Mamin, S. Chiang, H. Birk, P.H. Guethner, D. Rugar: J. Vac. Sci. Technol. B **9**, 1398–1402 (1991)
55. K. Matsumoto, M. Ishii, K. Segawa: J. Vac. Sci. Technol. B **14**, 1331–1335 (1996)
56. J.A. Dagata, J. Schneir, H.H. Harary, C.J. Evans, M.T. Postek, J. Bennett: Appl. Phys. Lett. **56**, 2001–2003 (1990)
57. E.S. Snow, P.M. Campbell, B.V. Shanabrook: Appl. Phys. Lett. **63**, 3488–3490 (1993)
58. Y. Okada, S. Amano, M. Kawabe, B.N. Shimbo, J.S. Harris Jr.: J. Appl. Phys. **83**, 1844–1847 (1998)
59. S. Sasa, T. Ikeda, C. Dohno, M. Inoue: Jpn. J. Appl. Phys. **36**, 4065–4067 (1997)

Printing: Saladruck, Berlin
Binding: Stein+Lehmann, Berlin